MW00712387

Randi,

I am so sorry to hear about your accident - I wish you a speedy recovery.

It was wonderful having you in my classes - you are the type of student that professors love - you are bright, active in class, and you work hard. These traits will serve you well once you graduate.

Take care and please keep in touch.

[signature]

Call me if you need anything
980-777-2277
or
262-JR-WEST

Legislator Use of Communication Technology

Legislator Use of Communication Technology

The Critical Frequency Theory of Policy System Stability

Joe West

LEXINGTON BOOKS
Lanham • Boulder • New York • London

Published by Lexington Books
An imprint of The Rowman & Littlefield Publishing Group, Inc.
4501 Forbes Boulevard, Suite 200, Lanham, Maryland 20706
www.rowman.com

6 Tinworth Street, London SE11 5AL, United Kingdom

British Library Cataloguing in Publication Information Available

Library of Congress Cataloging-in-Publication Data Available

ISBN: 978-1-4985-6529-5 (cloth : alk. paper)
ISBN: 978-1-4985-6530-1 (electronic)

∞™ The paper used in this publication meets the minimum requirements of American National Standard for Information Sciences—Permanence of Paper for Printed Library Materials, ANSI/NISO Z39.48-1992.

Printed in the United States of America

Contents

List of Figures

List of Tables

Preface

I've heard say that when all you have for a tool is a hammer, everything looks like a nail. In this case, my long career as an electrical engineer was my hammer, and the relationship between communication technology and the policy process looked like a big fat nail ripe for the hitting. Upon leaving the field of electrical engineering after twenty-five years to pursue a Ph.D. in Public Policy, it did not take long before the correlations between the engineering concepts and social science concepts in the field of Public Policy became evident. One of my early papers during my first semester as a doctoral student covered systems theory in the social sciences. It was not until the second semester of my first year as a doctoral student that the proverbial light bulb came on for me in a big way: my doctoral cohort was required to read Baumgartner & Jones' seminal book, *Agendas and Instability in American Politics*. As an engineer, the importance of system stability and negative and positive feedback was very clear to me, and Frank Baumgartner and Bryan Jones were speaking my language when they discussed feedback and policy system stability.

I recall drawing a feedback loop on the whiteboard for the very first meeting of my doctoral committee as I explained the importance of communication technology in the policy process feedback loop. Thankfully, my dissertation committee guided me to a dissertation project that was a bit less ambitious. They suggested I examine how state legislators used communication technology, one of the first steps in determining the role of communication technology on the policy process. Roughly four years later, I defended my mixed method dissertation research on how Arizona legislators use communication technology (West, 2014) and off I went to my first job as a university professor. I began my second year as a professor at the University of North Carolina at Pembroke in 2015 by launching a somewhat ambitious research project to determine how state legislators in every state used and valued a wide range of communication technologies. This book is the result of that research project, and it brings together my understanding of feedback loop theory and the quantitative and qualitative research I have completed. One of the purposes of this book is to develop and present a theory on how communication technology influences the policy process in the United States. The theory is developed in chapter 1, data is presented and analyzed in chapters 2 through 7, and the theory is combined with the data and explored in chapter 8.

I suspect many readers will find this book substantially different than many other scholarly books. Readers are not asked to simply *believe* that the analyses support the theories in this book based on some abstract summary of the analyses or references to existing research and secondary data; the analytical models stem from original research and are presented in detail, giving the reader the information needed to determine for themselves, what results are important and what results are not. Such detail may present a problem for some readers who may find the extensive presentation of regression models and tables off-putting, but they are included for an important reason; providing detailed analytical results not only establishes a foundation for the theory developed, but also provides much greater insights for the scholars who will read this book in the future, and who may potentially build upon it, or refute it. Simply put, there is so much information stemming from this research that I cannot possibly predict all of the ways that this information may be useful to future researchers, so I err by providing more information rather than less. Moreover, rather than presenting only the statistically significant results, I present *all* of the results, even those that are not statistically significant. I am a fan of the belief that the lack of a statistically significant relationship is just as important as a statistically significant relationship, although perhaps a bit less glamorous in the eyes of researchers.

As an aid to understanding, every section in the chapters that present regression models contains an explanation of a small representative sample (typically two explanations per model) of the results shown in regression model summary tables. Readers can be comfortable jumping to any chapter of the book knowing that they do not need to remember how to interpret ordinal logistic regressions with odds ratio results, or how to interpret dy/dx marginal effects analyses post processing for dichotomous variables in logistic probit regressions.

In my experience as a consumer of scholarly books, I find it helpful when an author provides a brief road map of where a book is heading so that concepts can more easily be contextualized. This book is organized so that readers may quickly jump to the sections that they find interesting, and can use the provided analyses or ignore them and jump to the policy implications sections. From a 30,000-foot level, this book opens with a theoretical exploration of communication technology using feedback theories from electrical and control engineering perspectives. In subsequent chapters, the relationships between communication technology frequency of use and importance, and demographic, political, and institutional (DPI) variables are examined using data from a 2015 to 2016 study of state legislator communication technology attitudes and behaviors. These relationships are presented, and then, where possible, compared and contrasted with existing research to understand the policy process implications of legislator communication behaviors. In the final chapter, the theory presented in chapter 1 is used as a lens through which to view the data gathered for this book, and a theory of the impact of communication technology on the policy process in the United States is offered.

Acknowledgments

My sincere gratitude is extended to . . .

My wife, Dr. Tracy West; without her year-long support for this effort, this book would never have been written. She lost many nights and weekends with me while I was working on this project. Additionally, she proofread my dissertation, many of the original journal articles, and conference papers that ultimately contributed to the chapters in this book, a task which must have been painful for her.

My former research assistant and now University of North Carolina at Charlotte Doctoral Student Jingjing Gao. Jingjing labored over hundreds of pages of analytical data and transferred key information into many of the tables in this book, effectively allowing me to write while she completed the . . . ahem . . . less glamorous aspects of writing a book. Jingjing also assisted by proofreading the final manuscripts. Without her assistance, this book would not have been completed on time.

Dr. Robert Schneider at the University of North Carolina at Pembroke, for his unwavering encouragement and support to write this book. Dr. Schneider's guidance regarding the process of writing a book (he recently finished book number three) and his encouragement were instrumental in keeping me motivated and on task. Additionally, Dr. Schneider acted as a mentor when I had questions about the process of writing a book.

My dissertation committee at Arizona State University's School of Public Affairs: Dr. Elizabeth Corley, Dr. James Svara, and Dr. Erik Johnston. Many of the concepts in this book were shaped by my dissertation, and my dissertation was shaped by my dissertation committee. I appreciated their almost five years of guidance (and patience) as I progressed through my PhD in public policy and public administration. They should all be sainted for what I put them through.

My Associate Acquisitions Editor Emily Roderick and her Assistant Editors Courtney Morales and Alison Keefner. This team guided me through this process, answering all of my questions, and putting up with a draft manuscript, a portion of which was a week late. I write these words just as I am submitting this manuscript for review. I have a feeling that I will be thanking them even more for all the work they will do to make this book a reality.

Ms. Amelia Elk, my Administrative Assistant at the University of North Carolina at Pembroke. Amelia freed me from some of the tasks that a Program Director typically performs, allowing me more time to focus on my research and writing.

The University of North Carolina at Pembroke. I appreciate the university's support by giving me the time to write this book, and for hiring me to do the greatest job in the world: working as a university professor.

Chapter 1

Communication Technology as Feedback in the Policy Process

OVERVIEW

This book was written to accomplish three primary goals: (1) to present data and analyses that can be used by other researchers to further their field of study; (2) to create a general theory on the role that communication technology plays in influencing the policy process in the United States and to use data to support this theory; and (3) to provide relatively easy-to-understand analyses that can be used by laypersons in the field of political science.

Goal numbers (1) and (3) are relatively self-explanatory, but Goal number (2) requires a brief explanation. The theory in Goal number (2) can be summarized as follows: Communication technology is the conduit over which the policy environment[1] communicates the need for policy change. Disturbances in the policy environment (Baumgartner and Jones [1993] would call these disturbances "punctuations") are communicated to legislators in part, by constituents and their legislator peers. Disturbances can be attenuated or amplified by how constituents and legislators use and value the communication technology. Attenuation of feedback from constituents and peers (and the policy environment in general) effectively stabilizes the policymaking process, while increases in the frequency of policy feedback can destabilize the policymaking process. Each policymaking institution has an associated critical frequency (ω_c) of communications above which the policymaking institution becomes unstable and processes policy much more quickly than when in a stable (or negative feedback) mode. Put succinctly, when the number of communication events from the policy environment reaches some critical threshold, the policy process speeds up.

Chapter 1 lays the foundation for the theoretical structure of this book by examining the role that communication technology plays in the policymaking process from a natural science perspective. Using the physical characteristics inherent in all communication technologies, chapter 1 offers evidence that the choice of communication technology used by legislators and constituents to communicate policy-related information matters when it comes to creating policy. In some cases, the communication technology used to communicate the policy message matters more than the contents of the message itself. For example, as we will see in chapter 4, legislators note that boilerplate e-mail from constituents not only have no impact on how they will vote on a policy issue, the e-mails actually leave the legislator with a negative impression of the constituent sending the boilerplate e-mail. In addition to laying the foundation for understanding how the characteristics of a communication technology influence the policymaking process, chapter 1 also unites portions of several theories of the policy process. Laying the groundwork for chapter 1 requires an understanding of basic terminology used in policymaking theory and control systems theory. This process begins with a discussion of the linkages which connect legislators with each other, and with their constituents.

LINKAGES

For the purposes of this book, linkages are best defined as any technology (including face-to-face communications) used to communicate policy information between policymakers and policy stakeholders. The term "linkage" is widely used in political science, but one of the earliest definition of linkages from a political science perspective was penned by Jewell and Patterson (1966) who noted that linkages provided "a channel of communication between leaders of important social groups and the government" (10). In this book, legislator to constituent and legislator to legislator linkages are examined to determine the potential impact of communication technology frequency of use and importance of the policymaking process. Because communication technology use is expected to vary as a function of legislator behavioral roles, it is important to understand why legislators communicate.

WHY LEGISLATORS COMMUNICATE

Scholars typically note that legislators seek and maintain their office driven by three primary influences or goals: (1) the need for reelection (Arnold, 1992; Fenno, 1973; Mayhew, 2004; Rohr, 1986; Wahlke, 1962). In essence, legislators communicate because they want to be reelected, regardless of whether their motivations for being reelected are altruistic (Arnold, 1992; Campbell, 2003; Rohr, 1986; Sabatier, 2007; Schneider and Ingram, 1997;

Wahlke, 1962) or self-interested (Cox and McCubbins, 2007; Forgette, Garner, and Winkle, 2009; Herrick and Thomas, 2005; Schneider and Ingram, 1997); (2) to make and enact policy (Fenno, 1973; Mayhew, 2004; Oleszek, 2011; Rosenthal, 1997; Sabatier, 2007); and (3) to acquire and maintain political influence or power (Campbell, 1982; Canfield-Davis, Jain, Wattam, McMurtry, and Johnson, 2010; Kathlene, 1994; Miller and Stokes, 1963; Wyrick, 1991).

Reelection is typically viewed as the most important of these goals, and interactions (linkages precipitated by communication technology as broadly defined in this book) with their constituents are necessary to achieve this goal. Early (mostly pre-Internet[2] era) researchers suggest that linkages between the various legislative roles and constituent relationships were at best weak (Arnold, 1992; Hedlund and Friesema, 1972; Kingdon and Thurber, 2003; Miller and Stokes, 1963). As we will see, the data gathered and examined for this book supports this contention. Armed with this precept regarding why communication is important to legislators, the next step is to provide a brief examination of legislator roles which are associated with why legislators communicate.

LEGISLATOR ROLES

Edmund Burke (1891) broke representation focus into two foci for legislators: (1) local interests and (2) national interests. As the opening quote to this section infers, Burke viewed legislators primarily as Trustees and viewed constituents as unenlightened. Eulau et al. (1959) expanded the Burkean roles to include three variations: (1) the Delegate, (2) the Trustee, and (3) the Politico. Eulau described these roles as follows: (1) the Delegate as a role in which the legislator believes that they are bound by the will of their constituents and must act as an arm of their constituents, (2) the Trustee as a role in which the legislator is bound by their own will and expertise, and (3) the Politico as a hybrid role of the first two in which the legislator is bound to make decisions based on the political situation, sometimes acting as a Trustee and sometimes as a Delegate as the situation warrants. In ideal forms, the Trustee has no need to communicate with constituents, while the Delegate and the Trustee-Delegate hybrid Politico depend on communications with constituents.

In *Rethinking Representation*, Mansbridge (2003) extends earlier work on legislator roles by presenting four forms of representation: (1) Promissory, (2) Anticipatory, (3) Gyroscopic, and (4) Surrogate. Each of Mansbridge's legislator roles requires a form of communication between constituents and legislators (and in the case of the surrogate, *non-constituents* communicate with legislators who champion a specific policy). A much more extensive review of legislator roles is covered in chapter 6. For now, it is sufficient to note that legislator roles are related to communication behaviors and motivations. Armed with a basic understanding of linkages, why legislators communicate, and legislator roles, it is time to focus on some basic definitions necessary to understand the underlying theory presented in this chapter.

DEFINITIONS

Communication Technology

In the simplest terms, communication technology is the physical medium through which policy feedback is transmitted. No matter *who* is communicating feedback regarding a particular policy: constituents, legislators, bureaucrats, lobbyists, or other stakeholders, some form of communication technology (as defined herein) must be used. For the purposes of this book, it is useful to broadly categorize communication technology into three distinct typologies based on their characteristics: (1) mature[3] communication technologies, (2) Internet-enabled[4] communication technologies (IECT), and (3) mass-media[5] communication technologies.

Some readers may balk at the use of the term "communication technology" for certain types of communications. For example, some may find it odd that face-to-face communications are categorized as a communication technology. There are a number of reasons why this was done: First, technology is frequently used for what we might refer to as face-to-face communications. For example, using WebEx™ or Skype™ allows people communicating to see body language and hear important intonations that convey information. Second, from a theoretical perspective, it is more interesting to analyze the most common forms of communication used[6] by legislators and constituents, to separate them into specific categories, and then to compare and contrast these categories. For example, how do legislator behaviors associated with face-to-face communications compare to legislator behaviors associated with e-mail communications? Finally, the use of the term "communication technology" allows future readers of this book to think in terms of the forms of some future communication technology available to them were neither available, nor possible to conceive of, at the time this book was written. For example, in the future, some form of virtual reality communication may exist that mimics *exactly* the face-to-face in-person communications we experience today, including senses such as touch and smell. Or, for a second example, perhaps a communication technology will be invented that, with a simple high technology implant, will allow humans to communicate with each other over long distances by simply "thinking" the communication. Sound far-fetched? It is easy to speculate that individuals one hundred years ago might have thought that long-distance communications via a watch (such as the third-generation Apple Watch™) were pretty far-fetched. Yet, here we are. With this brief introduction completed, the following section outlines some important definitions.

Constituent

Constituents are defined as the individuals who live in a legislator's defined district and are directly responsible for electing the legislator. Importantly, constituents are geographically bound to the legislator.

Peer

Legislator peers are defined as other legislators who are elected to the same local, state, or federal legislature. For example, the peers of North Carolina state legislators are other North Carolina state legislators (whether they be in the House or Senate), but Arizona state legislators would not be considered peers of North Carolina state legislators.

Frequency of Use

Frequency of use is defined as how frequently a legislator uses a particular communication technology. There are *no* limitations placed on this variable with respect to the purpose of the communication. For example, legislators were not told to respond to frequency of use for campaigning, or frequency of use for official business. They were simply told to indicate how frequently they used a communication technology overall, independent of the purpose of the communication.

Importance

Importance was defined for legislators as follows: "Importance is related to the likelihood that you will respond favorably to a request received from a peer (or constituent) via one of the communication technologies." Unlike frequency of use, importance was limited in scope to legislators' official duties.

Communication Technology Typologies

Communication technologies are categorized into three typologies: (1) mature communication technologies; (2) mass-media communication technologies; and (3) Internet-enabled communication technologies. Table 1.1 contains a listing of each of the individual communication technologies for each typology.

Table 1.1 Communication Technology Typologies

Mature	*Mass-Media*	*Internet-Enabled*
Face-to-face	Press (Newspaper/Magazine)	Text Messages
Telephone	Television	Twitter™
Hardcopy Letters	Radio	Facebook™
	Town-hall Meetings	YouTube™
		Blogs
		e-mail
		Web pages

Self-Generated by author using Microsoft Word.

Useful Control Engineering Definitions

Since this book uses terminology from both electrical engineering and control systems engineering, it is important to define these concepts early in the book since many political science–focused readers may not be familiar with these engineering definitions. The following control engineering definitions come from Ogata's *Modern Control Engineering* (Ogata, 1990):

Systems: A system is a combination of components that act together and perform a certain objective. A system is not limited to physical entities. The concept of the system can be applied to abstract, dynamic phenomena such as those encountered in economics.

Disturbances: A disturbance is a signal that tends to adversely affect the output of a system. If a disturbance is generated within the system, it is called *internal*, while an *external* disturbance is generated outside the system and is an input.

Feedback Control: Feedback control refers to an operation that, in the presence of disturbances, tends to reduce the difference between the output of a system and some reference input and that does so on the basis of this difference.

Closed-loop Control Systems: Feedback control systems are often referred to as *closed loop control systems*. In practice, the terms feedback control and closed-loop control are used interchangeably.

Open-loop Control Systems: Those systems in which the output has no effect on the control action are called *open-loop control systems*. In other words, in an open-loop control system, the output is neither measured nor fed back for comparison with the input.

Closed-loop versus Open-Loop Control Systems: An advantage of the closed-loop control system is the fact that the use of feedback makes the system response relatively insensitive to external disturbances and internal variations in system parameters (4–5).

Useful Electrical Engineering Definitions

The following electrical engineering concepts are discussed in this book. Definitions are derived from *Communication Systems Engineering* (Proakis, Salehi, Zhou, and Li, 1994):

Bandwidth: The range of frequencies or information that can be transmitted. Higher bandwidth corresponds to a larger frequency range and/or greater information content.

Latency: The delay between the time information or data is transmitted and the time it is received.

Duplex: Signal characteristics associated with transmitting information. Full duplex allows information to be transmitted to and from a receiver and transmitter at the same time, for example, a telephone call. Half duplex allows information to be transmitted to and from a receiver, but only one direction at a time, for example, a walkie-talkie. Simplex allows information to transmitted only one direction, from transmitter to receiver, for example, a television or broadcast radio signal.

Critical Frequency ω_c: ω_c is the value of value of ω where the loop phase of the feedback system is -180 degrees. ω_c is also sometimes referred to as the phase crossover frequency. In engineering terms, ω is defined to be 2π * (frequency in instances [cycles or bits] per second).

Frequency: Frequency is the inverse of time (F = 1/T). Frequency is measured in Hertz (cycles per second) for analog communications or Baud (bits per second) for digital communications.

With the important control and electrical engineering concepts defined, the theoretical structure for understanding the link between communication technology and policy process feedback can be examined. The next section offers a brief overview of how system level theories of the policy process have traditionally treated feedback.

SYSTEMS THEORY, FEEDBACK THEORY, AND THE POLICY PROCESS

Systems theory has long been a staple in simplifying a complex set of relationships so that they can be easily understood. As defined by one of the pioneers of general systems theory Von Bertalanffy (Ludwig, 1972) systems theory is defined by "a set of elements standing in interrelation among themselves and with the environment" (417). Systems theory can be thought of as a level of abstraction that is useful for creating a basic understanding of complicated concepts and relationships. Breaking down a complex set of concepts and relationships into smaller blocks and noting the interactions between the blocks can quickly lead to an understanding of how a system operates which is not easily obtainable through other analytic techniques. Systems theory is used in a wide array of applied sciences including the electrical, mechanical, biological, chemical, and aerospace engineering fields, and in economics, among others (Chen, 1989).

In social science theory, a systems approach has been less popular. This unpopularity stems, in part, from the unmet expectation that using natural science quantitative systems analysis techniques would lead to predictive answers to policy process problems (Stewart and Ayres, 2001). Nonetheless, systems approaches are still used in social science areas such as organizational theory (Denhardt, 2008; Katz and Kahn, 1978; Shirky, 2008), human behavior (Jurich and Myers-Bowman, 1998; Maslow, 1943; Skinner, 1965), and policymaking theory (Baumgartner and Jones, 2002; Kingdon, 1984; Sabatier and Jenkins-Smith, 1993).

In all closed-loop systems, feedback plays an important role that determines the stability of the system, whether the system is mechanical, electrical, or even theoretical (via computerized models). A relatively famous (for mechanical and civil engineers anyway) example is the Tacoma Narrows bridge, which was destroyed on November 7, 1940 due to harmonic feedback generated by the wind[7] interacting with the resonant frequency of the bridge structure, causing the structure to become open loop and be destroyed by the wind. YouTube™ has an excellent video filmed on the day the bridge was destroyed.

By way of a theoretical example of a closed-loop system, Baumgartner and Jones (1993) theorized the policymaking system in the United States as a "punctuated" equilibrium system where the normally slow and deliberate policy process in the United States (stability) is "punctuated" into speeding up due to some significant external feedback (instability). While Baumgartner and Jones[8] note that their theory is sometimes recognized as a theory of instability (256), they argue that their theory should be viewed with an equal focus on both stability and instability. This book follows their lead and focuses on both stability and instability, using communication technology as both the feedback path and a factor driving both stability and instability in the policymaking process.

In the following section, a brief review of basic feedback theory is presented. This discussion of feedback and systems theory is not meant to be a complete description of feedback in the policy process, but rather is an overview to be used as a tool to simplify the relationships between the policymaking process and communication technology by using control systems theory.

BASIC FEEDBACK THEORY

In terms of the policy process, feedback theory in its most fundamental form involves the system components shown in the basic feedback diagram in figure 1.1. The basic components are an input to the policymaking system, a summing node that combines the feedback signal β[9] with the input, a system function stage (A_{OL}) and an output. Among other uses, feedback can be used to stabilize a normally unstable open-loop system or to increase the stability of an already stable open-loop system (Ogata, 1990).

Translating figure 1.1 so that it focuses on policymaking systems, the input is the input to our policymaking block from the public policy environment, A_{OL} is the policymaking component (legislators and the institutions in which they operate), β is the policy feedback component from the public policy environment (constituents and other legislators), and the output is the policy that is designed to influence the public policy environment. T_1 is the policymaking block response time and is equivalent to the time it takes the policymaking block (legislators) to create a policy. T_2 is the time it takes for feedback to be processed by the public policy environment, T_3 is the latency from policy output to the public policy environment (legislators to constituents), and T_4 is the latency from the public policy environment to the policy input (constituents to legislators). The importance of these times will be discussed in the loop dynamics section toward the end of this chapter.

An Overview of the History of Feedback in System Level Theories of the Policy Process

Historically, the discussion of feedback in the policy process can be broken into three time periods: (1) Early (1960–1993) discussions of feedback in the policy process where feedback was noted as the connection from the

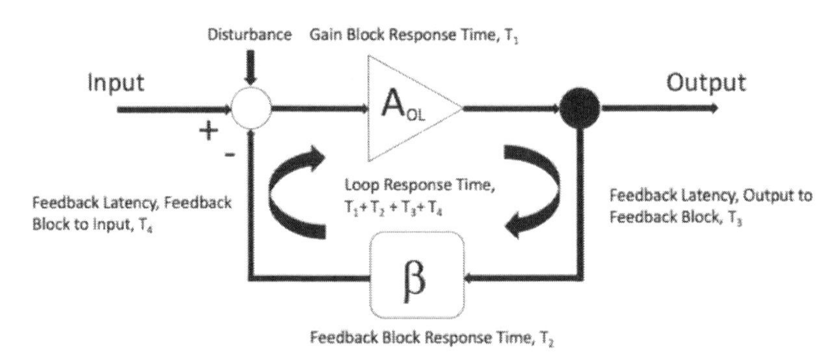

Figure 1.1 Basic single-variable feedback diagram.
Self-generated by author using Microsoft PowerPoint.

policy environment to the policymaking system. This period is defined by increasingly elaborate feedback paths, but little if any discussion regarding the implications of feedback characteristics on the policy process. Feedback is diagrammed in these early theories, but is not discussed in any meaningful way. (2) Pre-modern (1993–1997) defined by an increased focus on the effects of positive and negative feedback. This period was launched by Baumgartner and Jones (1993) seminal book *Agendas and Instability in American Politics*. In *Agendas and Instability in American Politics*, Baumgartner and Jones precipitated one of the earliest in-depth discussions of the relationships between the policy process, issue expansion, and positive and negative feedback. Baumgartner and Jones's work provides the conceptual impetus for the idea of using applied control and electrical engineering concepts as a lens through which to view the policy process in the United States. The control systems theory section of this chapter further elaborates on the important contributions of Baumgartner and Jones.

The modern time period (1997–present) is defined by more extensive focus "policy feedback effects." As defined by Soss et al. (2010), policy feedback effects emphasize "The potential for policies to play a causal role in politics" (14). These constructivist analyses place policy at the center of political transactions and serve as the foundation for examining political concepts. These concepts include the analyses of public policies such as social security (Campbell, 2003) and welfare reform (Soss and Schram, 2007), as well as policy concepts such as citizen engagement (Soss et al., 2010), power (Hayward, 2000), citizenship (Mettler, 1998), and justice (Shapiro, 1999). The study of feedback effects generally seeks to understand the relationships between public policies and the political process. As with earlier research on the policymaking process, this thread of academic research does not focus on understanding how the characteristics of the feedback path impact actual policy, which is the primary goal of much of this book.

At this point, there are two important concepts to remember: First, existing literature on policy feedback focuses primarily on the structure of the relationships between the public policy environment and the policymaking system *rather* than the characteristics of the feedback paths themselves. Second, as defined in this book, *every possible* feedback connection, from any point in a policymaking system to the public policy environment, and from any point in the public policy environment to any point in a policymaking system, is created by some form of communication technology. All policy feedback, regardless of where it is coming from and where it is going to, *must* be transmitted using some form of communication technology as defined in this book. With these two concepts noted, the next section examines the topic of policy feedback through an applied science, rather than a social science, lens.

VIEWING POLICY FEEDBACK THROUGH AN APPLIED SCIENCE LENS

With the notable exception of Baumgartner and Jones (1993), all of the policy process theorists discussed in the previous section treat the feedback process (β in figure 1.1) as an ancillary component of a policymaking system, choosing instead to focus on the policymaking components (congress, the public policy environment, constituents, special interest groups, etc.) rather than the characteristics of the policy feedback path. This is the point where the policy theory developed in this book diverges from all other known policy theories. Instead of treating feedback as a second- or third-order effect in policy process theory, this book focuses on the role that feedback plays in the policy process, first by disambiguating feedback into logical and understandable components based on applied science control systems theory and then by examining each type of communication technology using the appropriate characteristics such as bandwidth, latency, and duplex.

In this section, control systems feedback theory is used as a way to understand the various components of policy feedback that are important for developing a theory of the influence of communication technology on the policy process. The control systems engineering definition of feedback theory was chosen purposefully: *many* forms of today's communication technology, and *all* forms of Internet-enabled communication technology are designed using control systems and/or electrical engineering concepts. This fact makes these engineering concepts of feedback an ideal lens through which to view the impact of communication technology on the policy process. There are three distinct concepts that are important here. The first concept is that using an applied science control systems definition of feedback highlights the most important (first order, or primary) relationships between communication technology and the policy process. The second concept is that the electrical engineering characteristics associated with communication technology (bandwidth, latency, and duplex) will highlight less important (second order, or secondary) relationships between communication technology and the policy process. The third and final concept is that control systems feedback loop dynamics that impact system stability are directly applicable to policy process systems as theorized in this book.

Primary Characteristics

In the science of applied control systems engineering,[10] feedback has at least four important characteristics that are analogous to policy feedback characteristics from a communication technology perspective. These characteristics are: (1) frequency, (2) phase, (3) duration, and (4) path. Feedback in the policy process has similar characteristics: (1) feedback volume (frequency of communications), (2) policy issue expansion or contraction (phase, including positive and negative feedback), (3) feedback duration (length of time feedback is received), and (4) feedback path dampening (attenuation) associated with the communication technology used to communicate feedback. Each of these characteristics is discussed in the following sections.

Frequency

In engineering terms, frequency measures how fast a signal is changing, and it is measured in cycles per second (Hertz) or bits per second (Baud) in the digital realm. The higher the frequency rate, the more information which is transmitted, all else equal. In the public policy realm, frequency of communications regarding a policy issue is an indicator of policy mobilization, and is one of the driving forces toward instability in the policy process (Baumgartner and Jones, 1993). Feedback frequency can be thought of as the total volume (quantity) of feedback being received by a policymaker (in our case, legislators) from those impacted by policy (in our case, constituents) and other policymakers (peer legislators). For example, the frequency of feedback from constituents received by a legislator could be measured in phone calls per hour, e-mails per day, and so on. Increased frequency of communications is associated with issue expansion, but is not the only indicator (Sabatier and Weible, 2014).

Frequency of communication about a policy issue is a clear indicator that an issue is expanding; the higher the frequency of feedback that a legislator receives on a policy issue (regardless of where that feedback originates), the more likely it is that an issue is expanding, and the more likely it is that the policymaking system is moving toward instability (Baumgartner and Jones, 1993) . In policy theory terms, the greater the volume of feedback, the greater the likelihood that a disturbance in the policy process will take place (Walgrave, Varone, and Dumont, 2006). The concept of frequency is explained further in the section on critical frequency and crossover phase, which is followed by a more in-depth discussion of issue expansion. In an approach similar to that proposed in this chapter, Baumgartner and Jones (1993) measured the frequency of congressional attention to specific policy issues, in part, to determine their relationships to policy stasis and policy instabilities (innovation). Baumgartner and Jones noted that increased frequency of congressional hearings was associated with "punctuations" or instabilities/innovation in the policymaking process.

Phase

A feedback signal has two primary characteristics: frequency (just discussed) and phase. In applied control systems engineering terms, the phase of a feedback signal is one of the primary determinants of the stability of the system. If the phase of the feedback loop is additive (positive phase at the summing node in figure 1.1) the system will be unstable, and if the phase is negative (subtractive phase at the summing node in figure 1.1) the system will be stable, all else equal.

Phase in the context of the policy process is a function of the institution being analyzed, while frequency is a function of the public policy environment. Each political institution such as local government, state legislatures, and the federal government, has different phase characteristics. Each institution, when stimulated by some critical frequency of communication from the public policy environment, will transition from a stable (slow and methodical) policymaking process to an unstable (quick and potentially unpredictable) policymaking process. The frequency where an institution transitions from stability to instability is known as the *critical frequency*, and the phase at which this happens is known as the *crossover phase*. Frequency and phase are related, but separate concepts.

Communication frequency is the stimulus from the public policy environment, while the crossover phase is a function of the institution being stimulated. For example, if the institution being examined is a small city government, the critical frequency from the public policy environment may be 400 communication events per unit time. City government, when stimulated by 400 communication events from the public policy environment, may undergo a crossover phase from negative to positive, precipitating policy process instability. On the other hand, if the institution being examined is a state legislature, the critical frequency may be 10,000 communication events per unit time from the public policy environment before reaching the crossover phase transition from stability to instability, and at the federal level, the critical frequency may be 100,000 communication events per unit time. The number of communication events from the public policy environment required to precipitate instability of a policymaking system can vary substantially based on the institution and on the policy being discussed.

Issue Expansion

Just like phase is one of the determinants of stability in an electronic circuit, issue expansion is recognized by political scientists as an indicator of stability in the policymaking system (Baumgartner and Jones, 1993; Birkland, 2005; Sabatier, 2007). Baumgartner and Jones (1993) were some of the earliest researchers to note the relationship between the stability of the policy process and issue expansion. In terms of the policy process, Baumgartner et al. asserted that positive feedback is associated with issue expansion and instability in the policy process, while negative feedback is associated with issue contraction and the stability of the policy process (17–18). Baumgartner and Jones note that the United States is normally in a state of equilibrium, indicative of a slow and deliberative policy process which is resistant to change (homeostatic). Homeostasis is associated with negative feedback and with periods of equilibrium (Bertalanffy, 1956) in the policy process, while periods of instability are associated with very quick and sometimes dramatic policy changes (Baumgartner and Jones, 2002; Richardson, 1991; Paul Sabatier and Weible, 2014).

One important aspect of issue expansion that remains relatively unexplored by policy process theorists is that the speed with which issues can expand or contract is directly related to the communication technology used to provide feedback. In effect, communication technology itself plays an important role in driving a policymaking system from a state of equilibrium to a state of punctuation (from stability to instability). Let's examine a couple of (once again, admittedly ridiculous) corner cases to highlight this point. If there is no way for the public policy environment (constituents, policy stakeholders, special interest groups, etc.) to communicate their policy preferences, policy issues would never be able to "expand" or grow *except* through the policy-maker's own understanding of the impact of the policy on the public policy environment. Without communication technology, the feedback loop from the public policy environment shown in figure 1.1 is broken. A representative democracy such as that in the United States will be inherently unstable without feedback from the public policy environment. A system designed to be self-correcting based on feedback from constituents (broadly defined) loses its ability to self-correct when the feedback signal is lost. Note that this would not be the case in certain types of autocratic governments, such as single party dictatorships, which can, and do, function without feedback from those impacted by the policy. North Korea's despotic dictatorship under Kim Jong-un (like that of his father Kim Jong-il and grandfather Kim il-Song) is an example of this phenomenon.

The second corner case is defined by instantaneous communications—where all parties interested in a particular issue would know about the issue instantaneously and are able to communicate their interests instantaneously. In this case, issue expansion can occur in virtually zero time. In effect, the inputs to the policymaking system could vary so quickly that the policymaking system itself cannot respond, again, resulting in instabilities in the policy process. Imagine a legislature where feedback is coming from constituents constantly and is varying so quickly that legislators cannot assimilate the inputs quick enough to make policy. In this case, the inputs to the system are moving too quickly for the system to respond, resulting in an information overload that would, in theory, gridlock the system, once again resulting in an unstable system. Now let's examine two more realistic cases.

In the eighteenth century, communication "technology" was limited to two forms: face-to-face communications and hardcopy written communications in the form of handwritten letters or news an information printed by a printing press. The speed at which an issue could expand was limited either by how fast the information could be printed and then delivered, or by how fast it could be discussed in face-to-face meetings. In a large city, individuals could receive the daily newspaper very quickly (the first daily newspaper in America began in the late 1600s[11]), while news between cities was limited by the speed of a horse (the transcontinental telegraph system was not completed until 1861.)[12] Putting this in terms of issue expansion and policy process stability, the transition from a period of stability to a period of instability was limited by the speed of the horse, the speed of the printing press,[13] the speed of word of mouth, or some combination thereof. The US policy process was designed in just such a world, and as we will see in chapter 4 which discusses Internet-enabled communication technologies, a system designed for much slower feedback may begin to break down (become unstable) when feedback happens faster than the system can handle. By way of an electrical engineering analogy, our system of policymaking (the electronic circuit) was designed for the relatively slow feedback frequency typical of the eighteenth century.

Let's examine a second realistic case, a much more modern one. The theoretical maximum speed of an electrical signal in a wired communication system is 3×10^8 meters/second. In reality, communications via the Internet are much slower than this, limited much more by hardware processing times and signal transition times that are in the order of 1×10^{-3} (milliseconds). In the twenty-first century, President Donald Trump can send information on an issue important to him to approximately 20 million people via Twitter™ in less than 100 milliseconds (give or take). Each of these 20 million people could, if they wanted, send this to all of their Twitter™ followers in less than a minute. Such exponential growth of an issue was not even fathomable in seventeenth-century America. A quick comparison between the seventeenth-century theoretical issue expansion rates and the twenty-first century theoretical issue expansion rates makes it clear: a policy issue today can expand very quickly, and certainly much more quickly than it could in the seventeenth century. The fact that issues can expand faster suggests that there may theoretically be more destabilizing events possible today than in the seventeenth century. The next two sections provide a more in-depth examination of negative (stabilizing) and positive (destabilizing) feedback in the US policy process.

Negative Feedback

Negative feedback (homeostasis) is recognized as playing a role in the formation of policy (Baumgartner and Jones, 1993; F. Baumgartner and Jones, 2002; Beland, 2010; Lindblom and Woodhouse, 1993; Sabatier, 2007). In the policymaking process, negative feedback is associated with periods of stability, a concept that parallels feedback theory in the applied sciences. Negative feedback occurs when the output from a policymaking system becomes attenuated as it transitions through feedback paths and back into the input of the policymaking system. In the case of negative feedback the initial policy disturbance is suppressed by the feedback paths such that no political disruptions occur (Baumgartner and Jones, 1993). Negative feedback should not be conflated with the concept of policy feedback which is "negative" from a normative perspective.

Negative feedback mechanisms abound in American politics (Baumgartner and Jones, 1993, 256) and institutions that control policy do not frequently lose control, promoting political stability through institutional stability. There are a number of institutional forces that lengthen policy response times through negative feedback. Examples include a successful policy monopoly (Sabatier, 2007, 159), federalism (Baumgartner and Jones, 1993), separation of powers (Rohr, 1986), and jurisdictional overlaps (Baumgartner and Jones, 1993; Kingdon and Thurber, 2003; Sabatier, 2007), all of which inhibit major changes by providing negative feedback (Sabatier, 2007, 162). Baumgartner and Jones (1993) reinforce Sabatier's claim that policy monopolies promote institutional stability by suggesting that federalism provides an important limit to positive feedback processes though differences in receptivity to policy proposals because the various levels of government have substantially different policy priorities (219).

Negative feedback through federalism can occur in a number of ways: it can occur naturally, or it can be forced. In an example of a naturally occurring negative feedback, federal programs can grow until there is sufficient (from a feedback frequency perspective) feedback to suppress growth (Kingdon and Thurber, 2003). For example, in the 1980s, occupational safety regulations through the Occupational Safety and Health Administration (OSHA) became so oppressive that groups provided sufficient feedback to force policy change (104). In an example of forced negative feedback through federalism, the Food and Drug Administration (FDA) requires the testing of new drugs and devices prior to release to the public. The FDA intentionally slows down the feedback process by requiring corporations to follow specific (and sometimes lengthy) research study procedures designed to detect problems with treatments under consideration. The FDA could, if it wished, simply allow corporations to release drugs without any screening, which would be quicker, but which could potentially endanger the lives of patients using these drugs.

In a second example of federally forced negative feedback, Campbell (2003) notes that the US welfare process generates negative feedback through a degrading and highly personal means-testing process which tends to capitalize on the low level of welfare participants participatory resources to help ensure low levels of participation. In essence, the potentially degrading means-testing process limits the number of welfare applicants, acting as an input attenuation, which is a form of negative feedback. Welfare applicants, understanding that they will need to release personal and potentially sensitive information about their finances, may choose to avoid applying for welfare payments, which inherently stabilizes the welfare system by limiting the number of applicants and/or recipients.

Positive Feedback

"One thing leads to another," "the rich get richer," "an escalation of conflict"—according to Baumgartner and Jones (1993), these common sayings provide evidence of a general awareness in American society of the concept of positive feedback. In the mid-1990s, this awareness of the effects of positive feedback carried over into policy theory where it received attention because of its relationship to instability; positive feedback and instability (termed "punctuation" in their book) go hand-in-hand, whether the topic is in the applied sciences or in the social sciences. Baumgartner and Jones explained feedback from a social science perspective exceptionally well for the purposes of this book:

> When a system is subject to negative feedback, an initial disturbance becomes smaller as it works its way through time. In positive feedback, small disturbances become amplified, causing major disruptions as they operate across time. (6)

A thorough discussion of positive feedback in policymaking systems demands focus in three fundamental areas: (1) conflation of the normative and systems definitions of positive and negative feedback, (2) the relationship of positive feedback to system instability, and (3) qualitative examples of positive feedback applied to policy systems. The following sections address these areas beginning with a discussion of the multiple definitions of positive feedback that are often conflated.

The fields of political science and policy theory abound with discussions of positive (and negative) feedback. The meaning of the term "positive feedback" generally falls into two categories when it appears in political science and policy theory literature. The first category is the normative definition of positive feedback, which generally means to provide written or verbal feedback of a positive (approval) nature from one individual to another, while negative feedback is associated with disapproval. Many authors use the normative definition of positive feedback when discussing organizational behavior (Denhardt, Denhardt, and Aristigueta, 2008; Moran, 2013; Mowrer, 1960; Nahavandi, 2009; Scott and Davis, 2007). For example, Denhardt et al.

(2008) note in a section on employee rewards, "Be generous in your positive feedback when the situation calls for it" (163). The second category is the systems or applied science definition of positive feedback. This definition of positive feedback is associated with instabilities in the policymaking system and issue expansion (Baumgartner and Jones, 1993; Birkland, 2005; Overholser, 2016; Sabatier, 2007). The normative and applied science definitions for positive (and negative) feedback are commonly conflated, both in academic literature and subsequently, in the minds of those who read the literature. Academics exacerbate the conflation of positive feedback definitions by using both definitions in a single article and not noting the change in meaning. Unless otherwise noted, this book focuses *only* on the systems definition of positive and negative feedback.

Positive feedback causes small inputs or disturbances to amplify their way through a system, resulting in major changes or disruptions (Baumgartner and Jones, 2002; Lathrop and Ruma, 2010; Richardson, 1991; Sabatier, 2007). Pierson (2000) notes that positive feedback processes are "of great social significance" (251) and that social scientists are starting to advance meticulous arguments regarding the causes and consequences of positive feedback. This chapter furthers our understanding the importance of positive feedback by (in part) examining the relationships between disturbances (issue expansion), communication technology, and positive feedback.

Positive feedback is not limited to interactions between citizens and policymakers or between states through diffusion. There are many positive feedback relationships in the policymaking process that exceed these limited definitions. For example, Baumgartner (1993) introduces one such example by examining the interactions between policy images and venue assignment: "Changes in policy images facilitate changes in venue assignment. Changes in venue then reinforce changes in image, leading to an interactive process characterized by positive feedbacks" (80).

A final example of positive feedback in the policymaking process involves the media. Under certain conditions, a policy attracts the attention of the mass media which causes other media take notice and attract still more attention, ultimately resulting in a positive feedback mechanism as more and more media attention garner exponential expansion of the media coverage. While this is clearly the exception rather than the rule (a vast majority of policies and/or policy changes do not attract media attention), it is nonetheless an excellent example of the media acting as a positive feedback mechanism. Importantly, several of the forms of media used by legislators are analyzed in chapters 2 through 4. The following forms of mass media are included: television, radio, press releases, and town-hall style meetings.

Recognizing positive feedback processes *after* they have occurred is relatively easy and this section has presented a few examples of positive feedback processes. What has proven more difficult is producing theory to predict these positive feedback processes before they begin. This difficulty is best described by Baumgartner and Jones (1993):

> The most critical problem is that the theory we have developed in this book, like any theory based on positive feedback and strong interaction effects, offers strong explanatory power but little predictive power. One can model the results of a positive-feedback process, but one often has no idea exactly when that process might begin. (307)

The problem of predicting events likely to precipitate positive feedback is not now, and may never be, possible to solve. Positive feedback occurs when an issue reaches a critical tipping point (the crossover frequency ω_c) where it expands without control as events and relationships push it increasingly into the consciousness of mass media, policymakers, policy stakeholders, special interest groups, and constituents. The problem with predictive models is that there are too many unknown variables associated with the beginning of a positive feedback event. Simply put, this problem is too complex to create a predictive model using today's analytical tools. For example, the aircraft assault on the World Trade Center on September 11, 2001 precipitated the creation of many policies and the Department of Homeland Security. In this case, a terrorist attack precipitated the positive feedback that caused US policy to change quickly. Another example is the Sandy Hook elementary school attack on December 14, 2011 which precipitated gun control measures in Connecticut and other states. These two events highlight the difficulty of predicting the tipping point where a policymaking institution goes unstable. There are literally an infinite number of event possibilities which could precipitate policymaking process instabilities. In order to predict periods of instability, we would need to predict the event or events which lead to the instability. This is, using current technology, impossible.

Duration

Feedback duration is simply the length of time that feedback is received on any particular issue. Some policy issues, such as a citizen referendum, are relatively short-lived, while others, like gun control and abortion, are extremely long-lived. There are two typical corner cases:[14] (1) positive feedback is typically associated with high frequency short duration event and (2) negative feedback is associated with low frequency long duration events. One example of a *low frequency long duration* positive feedback is the Social Security policy in the United States (Campbell, 2003). Signed into law in 1935 by Franklin Delano Roosevelt (FDR), it wasn't until the benefits became widespread in the late 1950s that seniors, empowered by the policy itself, began to exert influence on policymakers to continue the program. The power of seniors has grown ever since and social security, through positive feedback, has become a political third-rail to those politicians who dare to threaten it (11). According to Campbell (2003), groups that benefit from positive policy feedback have greater participatory capacity and use that new capacity to further preserve the policy. The second feedback corner case can be categorized as a *high frequency short duration* positive feedback disturbance. This type of disturbance occurs in the diffusion of policies across political systems (Baumgartner and Jones, 1993; Walker, 1969). A modern example of such diffusion is Arizona's SB1070 immigration policy which diffused to eighteen states (Miller, 2010) in a matter of two months.

As can be seen from the above two examples, frequency and duration of feedback *together* lead to instabilities in the policymaking system just as high frequency long duration feedback in a control system leads to instability. In a control system, the duration of feedback that determines instability is associated with the system response time. For the policy process, frequency plus time can be combined into a measure termed "feedback density." Both high frequency short duration events and low frequency long duration events can have the same density, and both can lead to policy change; the former leads to change in an unstable public policy environment, and the latter leads to change in a stable public policy environment. The two other corner cases, high frequency long duration events and low frequency short duration events, are interesting as well, but no examples could be found. High frequency long duration events are the densest, and might be expected to lead to some form of political revolution. Low frequency short duration events are the least dense, and could be expected not to lead to policy changes. In an approach similar to that used in this chapter, Baumgartner and Jones (1993) noted the importance of the duration of congressional hearings in relationship to policy stasis and policy instabilities (punctuations), noting that longer duration congressional hearings were associated with instabilities in the policymaking process.

The concept of communication duration is relatively straightforward, but was *not* measured by the survey instrument since the duration of any communication event is associated with a particular policy. In the survey instrument, legislators were not asked about specific policies, effectively eliminating the possibility of asking about communication event durations. Along with crossover phase discussed previously, communication durations will not be analyzed in subsequent chapters in this book.

Feedback Path Dampening

The importance of any particular communication technology to a legislator will help determine how much weight the legislator assigns to feedback received via that communication technology. This effect is similar to dampening in applied electrical engineering. In one common example of this phenomenon, a special interest group will ask its online members to send representatives an e-mail by filling out a form. The text of the e-mail is automatically filled out and the e-mail is sent in the "name" of the individual to a legislator. Legislators sometimes receive these identical e-mails by the hundreds or thousands. Research into this topic suggests that this "bulk" or "boilerplate" e-mail is often treated by legislators as spam, and will be discarded (Best and Krueger, 2004; Mergel, 2012; West, 2014). By discarding a particular communication technology such as this, the legislator in effect attenuates (or dampens) the issue expansion effects associated with the unused communication technology. This dampening increases the stability of policymaking systems.

If a legislator ignores a feedback signal from a constituent or peer, either partially or in total, then the net effect of that feedback in contributing to issue expansion (and policy system instability) is reduced. Any communication technology which is unused by a legislator but used by their constituents (or vice versa) is negative feedback; it attenuates or reduces the feedback signal. Any communication technology used by both a legislator and their constituent contributes to positive feedback. Successful communication between a legislator and their constituent increases the frequency of the feedback loop, while unsuccessful communication precipitated by a communication technology mismatch between legislator and constituent acts as negative feedback and decreases the frequency of the feedback loop. Table 1.2 summarizes the stability traits of the primary characteristics associated with feedback loops.

Secondary Characteristics

Unlike primary characteristics that are associated with feedback theory, secondary characteristics are associated with concepts related to the communication technology used in the feedback process. Secondary characteristics tend to be less important than primary characteristics, but just like primary characteristics, secondary characteristics have significance in both applied (natural) science and social science fields. The concepts of bandwidth, latency, and duplex are important in this book. Fortunately, these concepts are treated similarly enough in the natural and social sciences that the reader will likely see the connection between the applied science and the social science

Table 1.2 Primary Feedback Loop Characteristics

Control Systems Parameter	*Policy Process Parameter*	*Increases/Decreases Stability*
Frequency	Frequency of Feedback, ω_c	Higher Frequency Decreases Stability
Phase	Crossover Phase of the Institution	Stable Below ω_c Unstable Above ω_c
Duration	Duration of Feedback	For a Given Frequency, Longer Duration Decreases Stability
Feedback Path Attenuation	Communication Technology Mismatch Between Legislators and Constituents and Between Legislators and Other Legislators	Increases Stability

Self-Generated by author using Microsoft Word.

applications fairly quickly. In the following sections, these three concepts are discussed first from an applied science electrical engineering perspective and then from a social science human communications perspective.

Bandwidth

In electrical engineering, bandwidth is the range of signal frequencies that a communication system utilizes. Bandwidth is measured in terms of the engineering concept of frequency, which has the units of cycles per second (Hertz in the analog realm) or bits per second (baud in the digital realm). Bandwidth is directly related to the amount of information that can be sent per unit of time, regardless of whether the information has been digitized or not. Many readers may not be familiar with these concepts from an engineering perspective, but most readers have likely experienced both analog and digital bandwidth applications. The examples in the next paragraph should help make this concept clear.

In the analog realm, the bandwidth of an FM radio signal is approximately 15 kHz, where k is the common engineering notation for 10^3 and Hz (Hertz) is the engineering notation for cycles per second. A bandwidth of 15kHz represents 15,000 cycles of information per second. The bandwidth of an analog telephone is 3kHz. Those who are old enough to have used an analog phone and FM stereo will immediately recognize how differently these two systems sound. A telephone sounds tinny and not at all like a real human voice, while an FM stereo broadcast sounds very much like the human voice and human instruments of music. Bandwidth is the difference. The typical human ear can respond to frequencies as low as 20Hz and as high as 20kHz. For a communication to sound "normal" to the human ear, it should have at least the same frequency range (bandwidth) as the human ear: effectively 20kHz. A 3kHz telephone signal has only 15 percent of the bandwidth of the human ear, while an FM stereo system has 75 percent of the frequency range of the human ear. In the digital realm, an early acoustic modem could transmit around 300 bits/second (300Hz) and today's fiber optic communication cables are designed to reach bandwidths of 255×10^{12} Hz. Just like the human ear, the human eye has a bandwidth. The estimated bandwidth of the human eye is roughly 10^6 bits/second, so a 300 bit/second acoustic modem could not transmit video that would appear normal, whereas fiber optic communication cables can transmit video many times faster than the human eye can process the video with an audio range that equals or is greater than the range of human hearing.

The topic of bandwidth from a social science perspective is significantly more complex than from the engineering perspective. In the social sciences, bandwidth is closely related to the concepts of social presence, media naturalness, and media richness. The concept of communication technology bandwidth plays a key role in understanding the policy process implications of legislators who use (or not) any particular communication technology. According to Burke et al. (1999), bandwidth is defined as the "range of cues transmitted by the [communication] medium; a higher bandwidth medium transmits more types of cues than one with less bandwidth" (559).

Notice how close the social science concept of bandwidth as defined by Burke et al (1999) parallels the engineering concept. The higher the communication technology bandwidth, the greater the communication technology's social presence. Larger media bandwidths can transmit more information (including social presence cues) over any given period of time as compared to lower media bandwidths. For example, high bandwidth video provides many more verbal and non-verbal cues than does a low bandwidth e-mail. Evidence of this is as close as your computer. An e-mail that takes five minutes to read consumes approximately 33,000 bits of hard disk space on a computer, while a five-minute video with audio consumes approximately 110,000,000 bits of hard disk space on a computer (depending on video format). The video with audio over 300 times the bandwidth than the e-mail. Bandwidth is the reason that emotions are difficult to transmit via e-mail, and why emoticons exist to give the reader an indication of the emotions associated with the information being transmitted in the e-mail.

Social presence (Dennis and Kinney, 1998; Lengel and Daft, 1988; Lilleker and Koc-Michalska, 2013; Rourke, Anderson, Garrison, and Archer, 1999; Short, Williams, and Christie, 1976) is an important concept in media richness theory. Social presence is defined as "the ability of learners to project themselves socially and affectively into a community of inquiry" (Rourke et al., 1999, 1). Social presence is the extent to which a communicator *feels the presence* of an individual with whom they are communicating.

According to Short et al. (1976), various communication technologies differ in their ability to communicate both quantity and type of information in a fixed timeframe. Short suggested that higher bandwidth communication technologies are associated with increased social presence, while lower bandwidth media types are associated with lower social presence. According to Rourke et al. (Rourke et al., 1999), social presence as a concept has its roots in Wiener and Mehrabian's (1968) concept of *immediacy*, defined as "those communication behaviors that enhance closeness to and nonverbal interaction with another" (203). An example may make this concept clearer. Face-to-face communications between a legislator and constituent who are alone in the same room will have a high degree of social presence, while an e-mail from a constituent to a legislator will have low (comparatively) social presence. Research by West (2014) suggests that legislators are aware of the concept of social presence and bandwidth. Here is the text from an interview with an Arizona legislator:

> Well, face-to-face always gives me the benefit of knowing . . . reading their personalities or aura or whatever it is about them; how they react to my statements, much better than e-mail. On the phone you can do it a little bit, but you know, but face-to-face always seems to work better as far as being able to read their reactions to an issue. (127)

Media naturalness theory (DeRosa, Hantula, Kock, and D'Arcy, 2004; Ned Kock, 2001; Kock, 2005) is related to both bandwidth and social presence. Media naturalness theory suggests that the more face-to-face like a communication technology is, the more physiologically satisfying it is, the less ambiguous it is, and the more information it can transmit over a given period of time. Media naturalness theory suggests that humans are evolutionarily "wired" for face-to-face communications. With roughly 300,000 years of *Homo sapiens* evolution (as of the 2017 finding in Morocco of the oldest *Homo sapiens* fossils), it would be difficult not to accept the importance of face-to-face communications in human societies. This said, while humans think they are good at reading emotions via body language, they are, in fact, not very good at it at all (Scheflen, 1972). Regardless, face-to-face communications have the highest social presence of any communication technology examined in this book.

Latency

Latency is nothing more than the delay between when a signal is sent and when the signal is received. One common example of latency occurs when watching a television program where a reporter is interviewing someone via a satellite phone. The reporter asks the interviewee a question over the satellite phone, and there is a significant pause before the interviewee responds. This pause is latency. The reporter asks the question, but the audio and video signal (the digitized words of the question and the image of the interviewee) has to travel up to a satellite in space, and then back down from the satellite to the phone in the hands of the interviewee. It takes up to a few seconds for the signal to go up to the satellite, be processed, and be transmitted back down to the interviewee.

The concept of latency in the public policy process is the same as the engineering concept of latency. It is associated with a delay. With respect to communication technology examined in this book, latency is defined as the delay between the time a communication is sent and the time is received. For example, in face-to-face communications where both of the individuals communicating are in the same room, the latency is very small,[15] in the millisecond range. Let's examine a latency at the other end of the latency spectrum. In the March of 1861, the Pony Express delivered Abraham Lincoln's first inaugural address from Nebraska to California in seven days and seventeen hours (Andrews, 2016). The latency of this communication format is seven days and seventeen hours. For the purpose of further clarification, let's hypothesize for a moment that President Lincoln was sending his inaugural speech to Nebraska so that a trusted advisor there could review the speech and send back comments. Assuming the latency to deliver the speech was the same as the latency to send the commented speech back to President Lincoln, the round-trip latency for this communication would be thiry-five days, ten hours, not including the time it takes to modify the speech. As we will later see, latency is an extremely important component of issue expansion and therefor a component of the stability of the policy process.

Duplex

With respect to communications systems, duplex is a characteristic that describes the flow of information from a transmitter to a receiver, and vice versa. There are three common forms of duplex: (1) Full-duplex indicates that information can flow from a transmitter to a receiver and from a receiver to a transmitter at exactly the same time, in a manner similar to a conversation where both individuals are talking face to face in the same room. For example, in telephone calls, both individuals on the phone call can talk at the same time and be heard at the same time. (2) Half-duplex indicates that information can flow from a transmitter to a receiver and from a receiver to a transmitter, but *not* at the same time. For example, when using a walkie-talkie or a citizen's band (CB) radio, one person presses their transmit button and speaks, and then when this person is done transmitting, the person on the receiving end presses their transmit button and then speaks. If both individuals try to transmit at the same time, the transmission will fail and neither of them will hear anything. There is a third form of communication directionality called simplex. Simplex communications are unidirectional; signals go only from the transmitter to the receiver. One common example of a simplex communication system is a baby monitor. The baby monitor transmits sound to the receiver, but the receiver cannot (in general) transmit any sounds back to the baby monitor. Television and radio station signals are also simplex in nature, although this is changing as television sets are increasingly capable of being connected to the Internet.

Table 1.3 summarizes the characteristics of the communication technologies commonly in use by legislators and constituents in the United States today. These characteristics will be analyzed using linear and non-linear modeling techniques in chapters 2 through 4.

Loop Dynamics

In control systems engineering, feedback systems such as the one shown in figure 1.1 have loop dynamic characteristics that are constantly changing. One of these characteristics is known as the general stability criterion. The general stability criterion specifies that closed-loop systems have a *critical frequency* ω_c. Closed-loop systems are, by definition, stable at frequencies below ω_c and unstable at frequencies above ω_c. In a feedback system, the

Table 1.3 Engineering Characteristics of Common Communication Technologies

Communication Technology	*Engineering Characteristics*
Face-to-Face	High bandwidth, low latency, full-duplex
Phone Conversations	Low bandwidth, low latency, full-duplex
Non-electronic Written	Low bandwidth, high latency, half-duplex
E-mail	Low bandwidth, low latency, half-duplex
Web pages	Moderate bandwidth, low latency, half-duplex
Text Messages	Very low bandwidth, low latency, half-duplex
Blogs	Low to moderate bandwidth, low latency, half-duplex
Facebook™	Moderate to high bandwidth, low latency, half-duplex
YouTube™	High bandwidth, low latency, simplex
Twitter™	Low bandwidth, low latency, half-duplex
Press Release	Low bandwidth, moderate latency, simplex
Television	High bandwidth, variable latency depending on format, simplex
Radio	Low bandwidth, variable latency depending on format, simplex
Town-hall Meetings	High Bandwidth, low latency, full-duplex

Self-Generated by author using Microsoft Word.

critical frequency (sometimes called the *phase crossover frequency*) is a function of all aspects of the feedback system including any disturbances to the feedback loop, the nature of the feedback, and the nature of the gain block.

The loop response time is the total time it takes for a policy to be created and feedback regarding that policy to arrive back at the input to the policymaking block. In figure 1.1, the loop response time equals $T_1 + T_2 + T_3 + T_4$. A system disturbance is shown injected into the input and feedback summing node. This disturbance or "punctuation" can be precipitated by many events that are themselves not related, resulting in a disturbance to the system that may not be predictable (Baumgartner and Jones, 1993) with respect to time, frequency, or amplitude.

Loop Response Times

With respect to loop response times, there are eight identified in figure 1.2. T_1 is the time associated with legislators creating a policy. T_2 is the time (expressed as a latency) that constituents take to evaluate a policy before communicating their policy preferences to their legislator(s). T_3 is the latency associated with legislators' evaluation of policy feedback from the public policy environment before communicating with their peers. T_4 is the importance path loss between the public policy environment and the time the legislator receives feedback from the public policy environment. Importantly, T_4 is the feedback path loss from the public policy environment that is *not* associated with their constituents. For example, some of the policy feedback may occur in the form of an opinion editorial that the legislator may read in a newspaper.

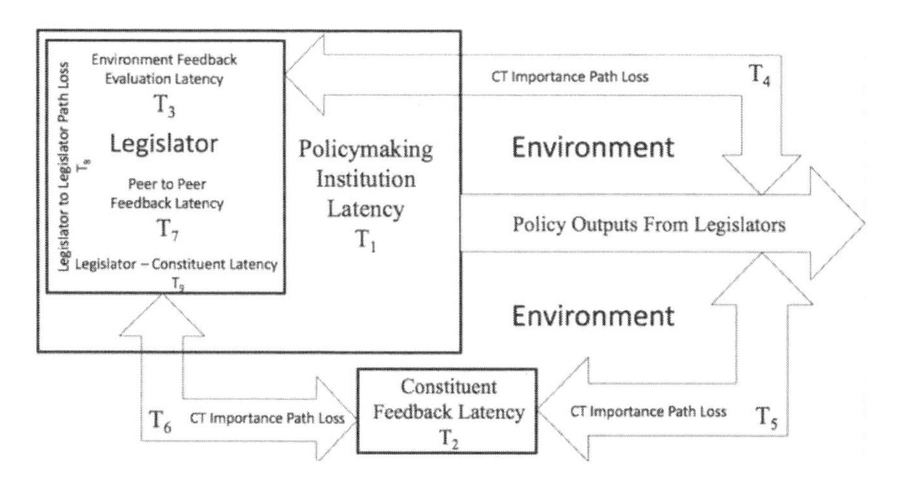

Figure 1.2 Policy feedback system latency and path loss diagram.
Self-generated by author using Microsoft PowerPoint.

If a particular legislator does not read or value newspapers, then this path loss would be 100 percent. T_5 is the importance path loss between the public policy environment and the constituent. By way of a second example; if the impact of a policy on the public policy environment is discussed extensively in the media (television, newspapers, radio, etc.), but a constituent does not read or value one form of media (radio for example), then the path loss for radio coverage of the policy would be 100 percent. Both T_4 and T_5 are the times associated with the media's attention to a particular policy (or a need for a policy) via some communication technology. T_4 and T_5 can most clearly be thought of as N-dimensional arrays representing a particular policy or event, a communication technology, a legislator (in the case of T_4), and/or a constituent (in the case of T_5).

T_6 is the path loss associated with the form of communication technology a constituent uses to convey policy feedback to their legislator. For example, a constituent may send his/her legislator an e-mail, but the legislator may not use e-mail[16] and therefore the path loss would be 100 percent. T_7 is the latency associated with a legislator processing information received by another legislator and T_8 is the path loss associated with the technologies used by legislators to communicate with each other. For example, if one legislator sends a text message to a second legislator, yet the second legislator does not read text messages, then the path loss is 100 percent. An example of a single feedback loop may make this concept clearer.

Let's say that a particular legislature is considering the passage of a law (policy A) that bans firearms outside the home. The policy topic is released to the public policy environment, and a newspaper B becomes interested in the policy topic and reports on the significance of policy A on the front page of their newspaper. Constituent C reads the newspaper (T_5), thinks about the article (T_2), and contacts legislator D regarding the issue (T_6). Legislator D considers the feedback from constituent C (T_9) and then communicates with legislator E (T_8) who then considers legislator D's position (T_7). In parallel with this sequence, both legislator D and legislator E are receiving information from the public policy environment (T_4) and are processing this information (T_3). Keep in mind that this process is running in parallel with every constituent interested in policy A, every legislator who will vote on policy A, and all forms of communication technology and media. In parallel with all of these events is the policymaking institution latency (T_1) that is considering the policy change. Now consider that this entire N by N dimensional array is duplicated for every policy under consideration by a state legislature, and is overlaid on a public policy environment that is constantly reacting to change and creating unpredictable disturbances that may be impinging on the policymaking process in many places all at once. The scope of this process should be clear; even from the standpoint of communication technology, the policymaking process is incredibly complex, and this complexity makes any model of the policy process an oversimplification of the highest magnitude.

With respect to loop response times and latencies in the policy process, policy theorists note certain aspects of policy process loop response times and latencies, but not others. For example, policy theorists note that the time for policymakers to enact policy, T_1, is shortened during periods of instability (periods of issue expansion) precipitated by disturbances (Baumgartner and Jones, 1993). The times T_4 and T_5 are also addressed by scholars who note that media attention to policy issues increases during periods of instability. This increase in exposure precipitates increased communications from many decision-making units (Baumgartner and Jones, 1993; Kingdon and Thurber, 2003), which include constituents and legislators. Note that the policy feedback diagram shown in figure 1.2 depicts the policy feedback system layered on top of the public policy environment and disturbances (which emanate from the public policy environment) can impinge at any point in the policy feedback system. One way to think of figure 1.2 is to consider the flow diagram as laying on a plane that represents the public policy environment. The public policy environment impinges on *every* point in the flow diagram, and disturbances emanate from the public policy environment and also impinge on every point in the flow diagram.

Disturbances

For the purposes of this discussion, the disturbance shown in figure 1.2 can be considered an input to the loop that can enter from the public policy environment at any point in the loop. Disturbances emanate from the public policy environment and can enter the feedback loop beginning with legislators, constituents, or the media, which, for the purposes of figure 1.2, are considered part of the public policy environment. When there are no disturbances (during periods of policy stability) the feedback loop is stable at some nominal loop frequency. When disturbances begin, they precipitate the communication events in the feedback loop. The longer the disturbance, the more communication events are precipitated. The larger the number of disturbances, related or unrelated, the more communication events are precipitated. Putting this last sentence another way; disturbances from the public policy environment excite and sensitize policy feedback loops. At some point, the frequency of communication events in the feedback loop exceeds the ω_c of the institution, precipitating phase crossover and subsequent instability in the policymaking process begins.

The frequency of feedback received by legislators from the public policy environment is commonly associated with media attention to a particular issue (Baumgartner and Jones, 1993; Paul Sabatier and Weible, 2014; Schattschneider, 1975). This media attention helps drive issue expansion. When media attention reaches a critical tipping point that varies as a function of the policy issue *and* the nature of the policymaking institution, communications from the public policy environment to the policymaking system reach a critical frequency, and the feedback loop is stimulated into instability. Disturbances can be precipitated by policy entrepreneurs (Baumgartner and Jones, 1993; Kingdon, 1984; Worsham, 2006), crises (Paul Sabatier and Weible, 2014), and other "focusing" events (Kingdon and Thurber, 2003, 94). An important characteristic is that *all* of these disturbances share one trait; they excite the frequency of the feedback loop until it reaches some critical frequency ω_c.

Critical Frequency/Crossover Phase (ω_c)

When a disturbance is introduced into the policy process system, on either a "wave" of popular (policy) enthusiasm (positive feedback from a normative perspective) or a "wave" of (policy) criticism (negative feedback from a normative perspective) (Baumgartner and Jones, 1993, 84), the policy process can transition into an unstable mode. Note that Baumgartner and Jones specifically use the term "wave." The term "wave" connotes an increased volume of enthusiasm or criticism, and that wave is measured in terms of increased communications. Increases in communication *frequency* (communicating support for, or criticism against, a policy) precede periods of instability in the policy process.

Tying this concept back to the policy feedback system shown in figure 1.2, this increase in communication frequency is injected into the policy process via the disturbance input (remember that this is a simplification; in reality, disturbances "blanket" the policymaking system, injecting their impacts over constituents and legislators alike), increasing the loop frequency until it reaches the critical frequency (the crossover phase) ω_c, at which point the policymaking system becomes unstable. The disturbances that cause this instability are not predictable (Baumgartner and Jones, 1993; Birkland, 2005), and the actual critical frequency/crossover phase is a function of the policymaking entity (e.g., local, state, or federal institutions). Increased communications are a necessary, and perhaps sufficient, condition to precipitate policy process instabilities. To summarize; the policy process in the United States remains stable until some external disturbance increases the frequency of communications from the public policy environment to policymakers until it reaches ω_c. Once the frequency reaches ω_c, the policymaking system transitions from stability to instability.

Because feedback phase is a function of the legislative institution receiving the feedback, feedback phase *is not* analyzed in this book. Remember, feedback phase or crossover phase is the frequency of feedback received (and processed) by legislators that causes the policymaking processes to go unstable for that institution. The steps to analyze feedback phase for each institution are relatively simple, but extremely time consuming. The possible steps to analyze feedback phase for a legislative institution would (as I envision it) consist of the following steps: (1) using archival data, looking back in time for the institution in question, identify policies that were produced more quickly than the average policies for the institution; (2) examine the frequency of communications from all possible communication technologies for a policy identified in item (1) above; (3) sum (or estimate) the total of all of the communication events (e.g., one e-mail would be one communication event) for every communication technology identified in item (2) above for the policy identified in item 1 above; (4) repeat steps (2) and (3) for each policy identified in step 1; and (5) average all of the policy frequencies (each of the step [3] numbers) to calculate the crossover phase for that institution. Repeat this process for every legislative institution of interest. At this point, it is likely obvious that the process of calculating the crossover phase for even one institution would be a significant task. Since this work was not completed as part of the data collection for this book, analyzing crossover phase is left for future researchers who find this concept sufficiently intriguing to pursue.

Figure 1.2 focuses on the policymaking institution. For the purposes of the research in this book, the policymaking institutions are state legislatures, although the concepts presented in this chapter apply to all policymaking institutions in the United States, regardless of whether they are at the municipal level or the federal level. The policymaking institution produces policy and releases it to the public policy environment. The public policy environment surrounds the entire policymaking system diagram and can enter at any point in the process. For simplicity, disturbances are shown at the input node, but they too surround the entire policymaking system diagram and can enter at any point.

Path Losses

There are four communication technology importance path losses shown in the loop pictured in figure 1.2. The first path loss is associated with legislator to legislator (peer) communications. For communication technologies that a legislator does not use and/or does not find important when communicating with other legislators, this path will be broken and the legislator to legislator feedback path will be open. The second path loss is associated with legislator to constituent communications, and, just like legislator to legislator communications, if there is a communication technology use mismatch, this communication feedback path will be open. The third and fourth communication path losses are associated with communication between the public policy environment and legislators and constituents. If a legislator or constituent does not use a communication technology being used by the media to report information on the public policy environment, the feedback loops between the public policy environment and the legislator and the public policy public policy environment and the constituent are open, effectively breaking the feedback loop.

Path losses, regardless of where they occur in the feedback loop are examples of negative feedback. Irrespective of the communication technology used, the choice of communication technology cannot amplify a feedback signal. By definition, path losses can only weaken the signal, and therefore can only act as negative feedback. An example may help clear up this concept. Let's say that the World Trade Center plane crashes acted as a disturbance to the federal policymaking system in 2011. A constituent with no Internet connection, no television, no radio, no print media subscriptions, and no neighbors might be completely unaware of this disturbance. The path losses in this case would completely open the feedback loop with respect to *this* constituent. The constituent could not increase the number of communication events received by their federal legislators, because they would not be aware of the disturbance. Their connection to the media is open so they can have no impact on the instability of the policymaking system; their lack of communication with outside information sources can only serve to stabilize the policymaking system. While this is an extreme example used to make a point, communication technology can be thought of as an information connection which can be open or shorted depending on whether or not a particular form of communication technology is used. The connection is open if the communication technology is unused and shorted if the communication technology is used.

The previous example treated communication technology as an open or short, but this is not to say that all communication technologies will stimulate the same increase in loop frequency. The effectiveness of a communication technology at stimulating feedback is not the same for all communication technologies. The ability of a communication technology to precipitate a communication event, effectively increasing the feedback loop frequency and moving the loop toward instability, is a function of the primary and secondary characteristics of the communication technology. The secondary characteristics of a communication technology, bandwidth, latency, and duplex, help determine how likely the communication technology will stimulate a communication event and increase the feedback loop frequency. Communication technologies with greater social presence are more likely to stimulate a communication event. This is to say that higher bandwidth, more face-to-face like, lower latency, and full-duplex communication technologies are more likely to precipitate a communication event. For example, a face-to-face conversation with a neighbor who lost a family member in the World Trade Center plane crashes is more likely to stimulate a constituent to communicate with their legislator than a generic tweet about the crashes. Figure 1.3 is a feedback diagram highlighting the important feedback paths associated with communication technology use. The usefulness of any particular communication technology to stimulate a communication event is based not only on the characteristics of the communication technology itself, but also on where in the feedback loop this communication event is happening. For example, a legislator tweet may be enough to stimulate feedback from a legislator to their constituents, but a similar tweet from a special interest group to a constituent may not be sufficient to stimulate the constituent to contact their legislator.

Research Design Overview

Dominating the choice of a research design for the study underpinning this book was the exploratory nature of the subject under study. This is *not* to

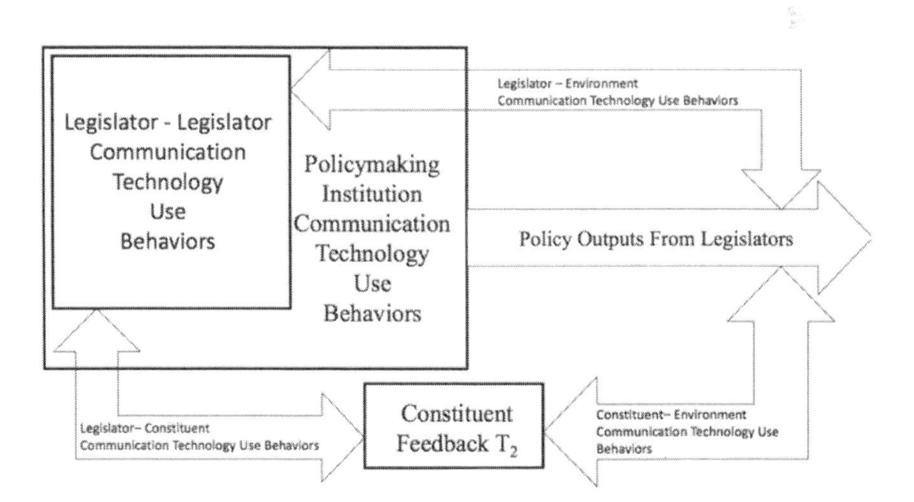

Figure 1.3 Policy feedback system use behavior feedback diagram.
Self-generated by author using Microsoft PowerPoint.

say that the topic of legislator use of communication technologies is unexplored, but rather, the data necessary to examine the research questions that intrigue me had not yet been gathered. Exploratory studies are traditionally used for three primary purposes, each of which is important for this study: (1) to satisfy a researcher's need to understand a phenomenon (Babbie, 2010; Lawrence, 2011; Sue and Ritter, 2011), (2) to test the waters for a more extensive study (Babbie, 2010; Lawrence, 2011), and (3) to develop the methods that can be used in subsequent studies (Babbie, 2010; Lawrence, 2011). Babbie and others note that exploratory studies are particularly appropriate for research that investigates previously unexplored topics (Babbie, 2010; Patton, 1990). After careful review of many research designs, a two-phase qualitative first, sequential exploratory mixed methods design was deemed optimal and selected.

As noted by Creswell and Clark (2007) exploratory mixed methods research is particularly useful when there are insufficient theoretical constructs with which to guide a study, when instruments and measures are unavailable to the researcher, and when a researcher wishes to develop an emergent theory. Each of these aspects reflects the conditions surrounding the research in this study at the time it was developed and conducted. One of the benefits of a sequential exploratory mixed methods approach is that a sequential approach allows a dynamic reconfiguration of the study focus based on participant responses, effectively allowing the study to refocus on unanticipated or important results.

QUANTITATIVE RESEARCH

Research Population

The research population for the data presented in this book focused on upper (Senate) and lower (House of Representatives)[17] legislative chambers in all fifty states in the United States, with the exception of the Nebraska legislature, which is unicameral in nature (Nebraska's legislators were coded as Representatives). At the time this book was written, there were approximately 7,383 state legislators in the United States.

Survey Modes

The survey data collection for the study referenced in this book consisted of Internet and mail survey delivery modes, modeled after the Tailored Design

Method approach developed by Dillman et al. (2009). Legislator e-mail addresses were obtained from state government websites for all states except Kentucky, New Jersey, and South Carolina. Because these three states do not publish their legislator e-mail addresses, researchers had to contact state officials to determine e-mail addresses. Survey links were e-mailed to all 7,383 legislators. Of the 7,383 e-mails sent, 1,421 e-mails were returned as undelivered due to e-mail address errors[18] and 988 were blocked as spam by state legislature information technology departments. The best estimate is that a total of 4,974 e-mails were delivered to state legislator e-mail inboxes.[19] Under the assumption that 4,974 e-mails were delivered to legislators and there were 1,056 responses, the *Internet* mode response rate for this study is 21.2 percent, a rate that is similar to the response rate for surveys conducted by other researchers surveying state legislators (Dietrich, Lasley, Mondak, Remmel, and Turner, 2012; Harden, 2011), but lower than other researchers who had response rates from legislators in the 40 percent to 60 percent range (Butler and Broockman, 2011; Carey, Niemi, Powell, and Moncrief, 2006).

As part of the Internet survey, legislators were invited to use an identifying alphanumeric code that enabled researchers to identify legislators who responded to the Internet survey so that mail surveys would not be sent to legislators who had already responded to the survey via e-mail.[20] A total of 684 out of 1,056 (64.8 percent) of legislators responding to the Internet survey included their unique alphanumeric code. Follow-up (reminder) e-mails were sent to all 4,974 legislators every seven days from the initial e-mail. Follow-up e-mails ceased on August 17, 2015, when legislator responses dwindled to fewer than ten per week, although the Internet survey was left open so that legislators could respond if they wished.

The second mode, the mail survey, began on September 10, 2015 and mail mode surveys collection was concluded on April 1, 2016.[21] Mail surveys were identical in format to the online survey, with the following exceptions: (1) they were printed instead of being displayed on a monitor and (2) they were not displayed one question at a time. The mail mode response rate was (691 mail mode responses from legislators)/(7,383−1,056 (Internet responses) − (187 letters which were returned as undeliverable)) = 691/(6,140) = 11.25 percent. The overall response rate for this study is (1,747 responses)/(7,383 total legislators−187 returned letters[22]) = 24.3 percent. Data collection intentionally spanned two legislative sessions: the fall session 2015 and the spring session 2016. Table 1.4 provides a detailed summary of legislator response demographics for the data that will be referenced in all subsequent chapters. The histogram shown in figure 1.4 highlights the number of responses by state.

Table 1.4 Legislator Response Demographics

Demographic Variable	*Number of Responses*	*Summary Statistics*
House of Representatives	1,213	72.77%
Senate	454	27.23%
Strongly Progressive Democrat	183	11.28%
Moderately Progressive Democrat	339	20.89%
Slightly Progressive Democrat	75	4.62%
Independent Leaning Democrat	98	6.04%
Independent	14	0.86%
Independent Leaning Republican	62	3.82%
Slightly Conservative Republican	44	2.71%
Moderately Conservative Republican	426	26.25%
Strongly Conservative Republican	382	23.54%
Other	0	0%
Male	1,113	73.64%
Female	389	25.90%
Years in Office	1,559	Min = 1 Max = 66 Mean = 7.91 σ = 8.04
Age	1435	Min = 21 Max = 88 Mean = 58.04 σ = 12.01
Education	1483	Min = 10 Max = 23 Mean = 17.30 σ = 2.87
White	1256	76.45%
Hispanic	25	1.74%
Puerto Rican	7	0.47%
Cuban	3	0.23%
Spanish or Latino	10	0.68%
Negro, Black	64	3.90%
American Indian	13	0.88%
Asian Indian	3	0.23%
Filipino	7	0.45%
Japanese	9	0.53%
Korean	3	0.23%
Vietnamese	2	0.15%
Native Hawaiian	4	0.30%
Samoan	2	0.15%
Other Asian	3	0.23%

Self-Generated by author using Microsoft Word.

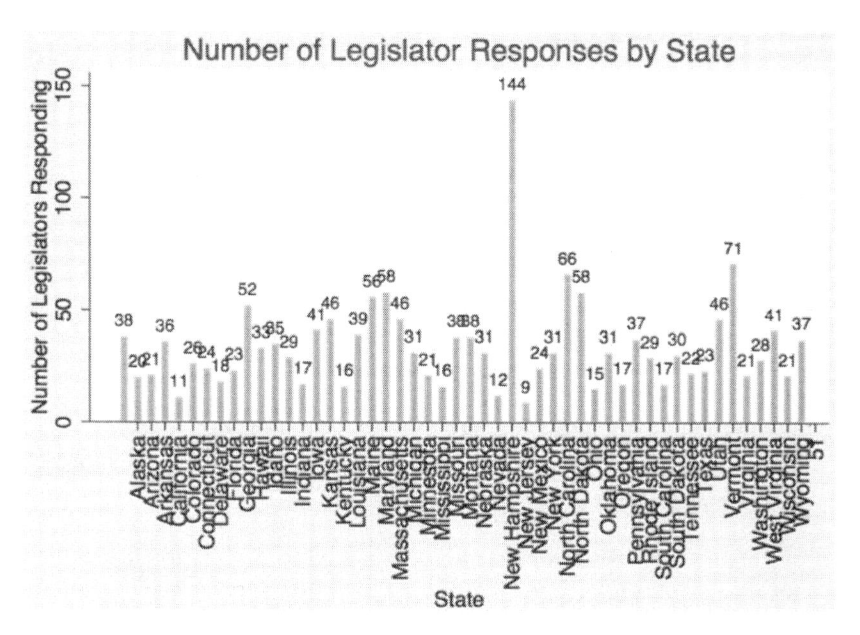

Figure 1.4 Legislator response demographics.
Self-generated by author using Stata Quantitative Analysis Program.

INTERNAL INSTRUMENT MEASURES

The survey consisted of an introductory cover letter briefly outlining the study and obtaining participant consent followed by seventeen questions. These questions consisted of nine demographic questions focused on the following variables: legislator age, gender, race (two questions), education, state, chamber, political party and strength of affiliation, and years in office. In addition, there were seven questions that were used to measure the dependent variables for the study. Specifically, there was one question focused on the *frequency of use* of communication technologies when used to communicate with *other legislators* and one question focusing on the *importance* of communication technologies when used to communicate with *other legislators.* These two questions were repeated by asking the same two questions with respect to legislators communicating with *their constituents.*

The questions regarding communication technology frequency of use and importance were asked multiple times across specific communication technology or hardware technologies. In particular, respondents were asked these questions about ten forms of communication technologies: (1) face-to-face meetings, (2) telephone calls, (3) letters (hardcopy), (4) e-mail, (5) Twitter™, (6) Facebook™, (7) web pages, (8) blogs, (9) YouTube™, and (10) text messaging. For constituent communications only, three additional forms of mass-media communications were included in the survey: (1) press releases, (2) television, and (3) radio. Legislator use of town-hall style meetings was also examined.

In addition to the four questions related to communication frequency of use and importance, there were three other dependent variable questions: One question examined the legislator's behavior as a Delegate (or Trustee or Politico); one question examined how frequently a legislator's policy preferences conflicted with the preferences of the majority of their constituents; one question examined how quickly the legislator responds to constituent requests. The final question examined how much time a legislator spent meeting with various individuals during a typical day: the list of choices included constituents, legislative staff, lobbyists and special interest groups, legislators from their own political party, legislators from other political parties, constituents from their own party, constituents from other political parties, legal counsel, government agency representatives, and constitutional officers (governor, attorney general, secretary of state, etc.).

In the survey instrument, the communication technology frequency of use variables is ordinal in nature, with the following response categories: do not use (coded as 0), use annually (coded as 1), use monthly (coded as 2), use weekly (coded as 3), use daily (coded as 4), and use hourly (coded as 5). The communication technology importance variables are also ordinal in nature with the following response categories: do not use (coded as 0), not important (coded as 1), slightly important (coded as 2), moderately important (coded as 3), important (coded as 4), and very important (coded as 5).

While the previous paragraphs discuss all of the variables in the survey, the research in this book also analyzes relevant information collected outside the survey. For example, state level gerrymandering information, citizen legislator or professional legislator status for each state, party majority/minority status, and term limits. These measures are discussed in the following section.

EXTERNAL INSTRUMENT MEASURES

Gerrymandering Measure

State level gerrymandering data was derived from the Azavea and Cicero (2012) whitepaper entitled "Redistricting the Nation: Redrawing the Map on Redistricting." Compactness scores for each state with a legislator that responded to the survey are coded as one of four variables based on the district compactness used and an overall district compactness measure that consisted of the sum of each compactness measure. Four measures of district compactness are used for all states that have more than one district: (1) state mean score of the Polsby-Popper compactness measure, (2) state mean score of the Schwartzberg compactness measure, (3) state mean score of the Convex Hull compactness measure, and (4) state mean score of the Reock compactness measure. Each of these independent measures gages district compactness on a scale where a lower score means LESS compactness (more gerrymandering) so the creation of a composite variable using the sum of each compactness score was appropriate. The variables associated with these scales included *pp*, *sch*, *con*, *reock*, and *gerry*. Additionally, a variable *compact* was used as a binary variable that represented all states that had laws mandating compact districts. Figure 1.5 shows the Maryland third district that has a low compactness score (high degree of gerrymandering) across all four compact measures.

The Polsby-Popper ratio and Schwartzberg ratio measure the perimeter and indentation of a district and measure how closely the district resembles a circle in order to determine compactness. The other two metrics

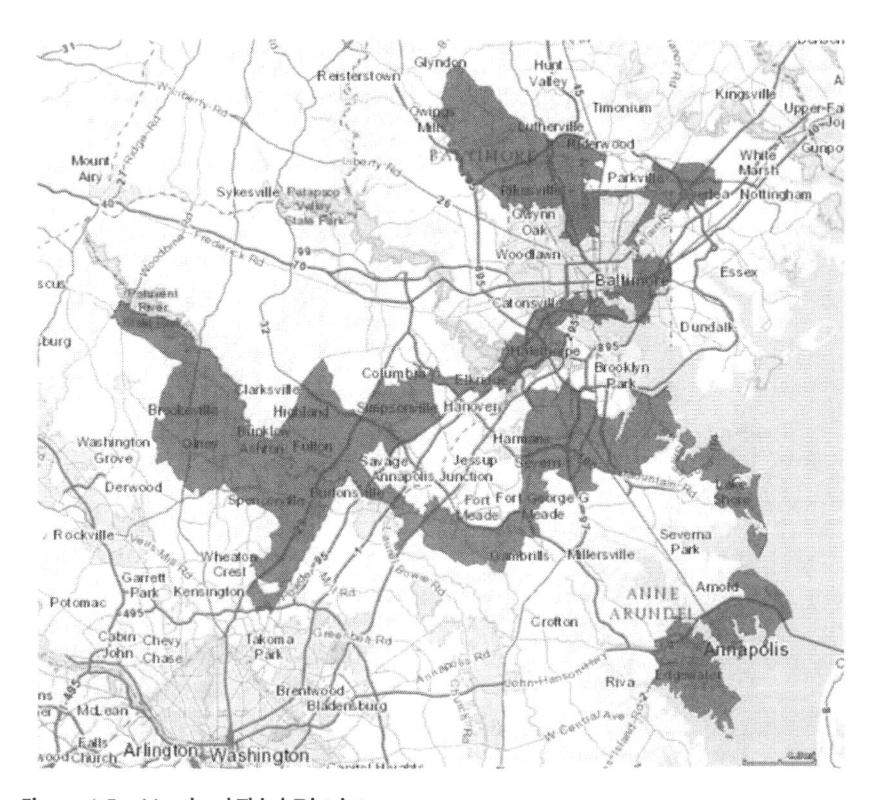

Figure 1.5 Maryland Third District.
Adapted from https://planning.maryland.gov.

Azavea used—the Convex Hull and Reock ratios—measure the dispersion of a district to assign a compactness score. The limitations on these measures as noted by Azavea and Cicero (2012) include the physical geographic boundaries of a state influence all four measures of compactness. For example, a state with relatively symmetric geographic boundaries such as Utah will have systemically higher compactness scores (meaning they will display less gerrymandering characteristics). States with less uniform boundaries such as Rhode Island will have more gerrymandering characteristics and lower compactness scores. Compactness scores were measured for 2012,[23] yet are compared against a 2016 measure of state legislators. This may cause a measure of error in the results reported in this chapter.

Party Majority/Minority Status Measure

Each state with a legislator who responded to the survey was analyzed to determine the minority party status of the state upper house (normally the Senate), lower house (normally the House of Representatives), and the governorship. State government web pages were used to determine minority party status. Several minority party variables were created: (1) *minboth* represents states where the majority party held both the House and Senate, (2) *minall* if the majority party held both houses and the governorship, and (3) *minparty* if the majority party held only one chamber (either the House or the Senate). A composite ordinal variable *minstatus* was created where 1 was coded if a political party held one chamber, 2 if a political party held two chambers, and 3 if a political party held both chambers and the governorship.

Professional Legislature Measure

Each state with a legislator who responded to the survey was identified as either a professional legislature or citizen legislature using a grading system developed by the National Conference of State Legislatures (NCSL). The variable *professional* was created and coded as follows: 0 if the state legislature was part-time, of low pay, with a small staff, 1 if the state legislature was part-time, of higher pay, larger staff, 2 if the state was classified as part-time but legislators spend more than two-thirds of a full-time job being legislators, 3 if the state was classified as full-time but legislators have lower pay and smaller staffs, and 4 if the state was classified as full-time, well paid, with large staffs. Additionally, the variable *professionalbin* was created which followed the traditional definition of a professional legislature. *professionalbin* was coded 0 if the legislature was part-time and 1 if the legislature was full-time.

Polarization Measure

Legislators were asked to identify their political party affiliation choosing from nine possible selections: Strongly Progressive Democrat, Moderately Progressive Democrat, Slightly Progressive Democrat, Independent Leaning Democrat, Independent, Independent Leaning Republican, Slightly Conservative Republican, Moderately Conservative Republican, and Strongly Conservative Republican. These options were coded as *partyid* from 1 to 9 respectively. Two polarization measures were created from *partyid*: (1) *polarized*, a variable coded 1 for independents, 2 for independent leaning Republican or Democrat, 3 for slightly (progressive or conservative) Democrat or Republican, 4 for moderately (progressive or conservative) Democrat or Republican, and 5 for strongly (progressive or conservative) Democrat or Republican, and (2) *polarizedbin*, a binary variable = 1 if legislator described their political party affiliation as strongly (progressive or conservative) Democrat or Republican, and 0 otherwise.

Term Limit Measure

The binary variable *termlimit* was created to measure the relationship between term limits and legislator role. *termlimit* = 0 if a state does not have term limits and *termlimit* = 1 for the states that have term limits.[24]

QUALITATIVE RESEARCH

Research Populations

Legislators

Face-to-face interviews with legislators were completed at the Arizona state capital. The Arizona state legislature is a bicameral body comprised of the House and Senate. The Arizona House (the lower body) is comprised of sixty members, two members from each of Arizona's thirty legislative districts. The Senate (the upper body) is comprised of thirty members, one from each legislative district. Arizona legislators are "citizen legislators," meaning that their positions are part-time in nature as opposed to professional legislators whose positions are full-time (with increases in salary, staff, and session length when compared to citizen legislators).

Legislator Assistants

Legislator assistants are Arizona state employees that are assigned to legislators in the House or Senate. Legislator assistants in both chambers serve "at the pleasure of"[25] the Speaker of the House or the President of the Senate. The pairing of legislator assistants and legislators differs slightly across chambers. Legislator assistants were selected for interviewing based on their pairing with interviewed legislators.

IT Department

IT department staff are Arizona state employees who report to the Arizona Legislative Council. The IT department consists of ten employees: one IT manager, one network administrator, one project manager, two programmers, one helpdesk supervisor, and four support specialists. IT department staff support the Arizona House and Senate, as well as the Legislative Council and Joint Legislative Budget Committee (JLBC). A total of three IT department staffers were interviewed.

CONCLUSION

Using control systems and electrical engineering concepts, this chapter has stepped through the process of creating the beginnings of a theoretical lens through which the effects of communication technology on the policy process can be viewed. This theoretical model suggests that there are links between communication technology and the stability of the policy process in the United States. These links in turn suggest that advances in communication technology can be expected to impact the policy process in three important ways: (1) the increasing speed of the communication links between the public policy environment and legislators can be expected to increase the frequency that policy process disturbances occur; (2) these higher speed communication technologies can be expected to increase the speed that policy issue expansion can occur; and (3) higher speed communications used by legislators in their institution may decrease the time it takes for a policy to be created.

In addition to these three effects, there are other subtler relationships between communication technology and the policy process. Fundamental concepts, derived from control systems theory, link communication technology with the policy feedback process. These links include the frequency of communication, the phase crossover point (the number of communication events before the policy process destabilizes) of the institution, the duration of communication between the public policy environment and legislators, and the path attenuation of the feedback precipitated by legislator and communication technology chosen by the public policy environment. Many of these primary concepts are associated with existing policy process theories.

Secondary concepts, derived from electrical and control engineering theory, link communication technology characteristics with the policy process. These communication technology characteristics include the bandwidth of the communication technology, the latency associated with the communication technology, and the duplex characteristics of the communication technology. Most of these secondary concepts have existing social science equivalents including media richness theory, media naturalness theory, and the concept of social presence.

Chapter 1 has presented the theoretical structures that will be used to drive the communication technology analyses completed in all subsequent chapters in this book. Moreover, this theory will be invoked at the end of each chapter, and in chapter 8 to tie these concepts together. Chapter 2 investigates the role that mature communication technologies play in legislator communications with their constituents and their peers.

NOTES

1. The policy environment is defined as follows: "The policy environment contains the features of the structural, social, political, and economic system in which public policymaking takes place" (Birkland, 2005, 201).
2. The Internet era is typically thought to have begun in the early 1990s, when Netscape created their web browser and use of the Internet began to grow exponentially.
3. Face-to-face meetings, phone conversations, and written/printed communications.
4. E-mail, Twitter, Facebook, YouTube, web pages, blogs, and text messaging.
5. Television, radio, press releases, and town-hall meetings.
6. It should be noted that scholars have used the term "linkage" to describe communications between legislators and constituents. The use of the term "linkage" by scholars is analogous to the use of the term "communication technology" in this book. Examples of the use of the term "linkage" make this analogy abundantly clear. For example, scholars have referred to linkages as talking and listening to constituents and reading their mail (Arnold, 1992; Kingdon and Thurber, 2003), newspaper columns, radio tapes, newsletters, opinion polls (Jewell, 1983), and phone calls (Keim and Zeithaml, 1986).
7. Or the positive feedback many readers have experienced which occurs when a microphone is too close to a speaker system.
8. A great deal of credit goes to Frank Baumgartner and Bryan Jones for their pioneering work which was instrumental in creating this book; as noted in the Acknowledgments, I am eternally grateful to them.
9. Not to be confused with the standardized regression coefficient ß.
10. The field in which I worked for approximately thirty-five years, which explains my choice of an applied science field on which to base this analogy.

11. Selnow, 1998.
12. Andrews, 2016.
13. Optical communication was used by the military since at least Roman times, but optical communications were not in widespread use in the United States until the twentieth century.
14. Corner cases are examples of extreme conditions leading to some outcome under investigation. Corner cases help show the "limits" of relationships and serve as "thought experiments" to facilitate understanding of the relationships under consideration.
15. It can be calculated as the distance between the speakers in meters divided by speed of sound (340.29 meters/second). The units in this case are meters/(meters per second) = seconds. If two speakers are one meter apart, the latency would be 1/340.29 = 2.94ms, basically, so fast that the delay would not be detected.
16. Ten legislators responded to the quantitative survey indicating that they do not use e-mail to communicate with constituents.
17. Lower chambers in certain states are known under different names. For example, in California, the lower chamber is referred to as the California State Assembly. For simplicity, in this study I will refer to all lower chambers as the House of Representatives.
18. Including closed e-mail accounts, errors in e-mail address coding, and errors in legislator contact information web pages.
19. State information technology departments can block e-mails "silently" with no errors sent back to the sender. It would be difficult to detect when this occurs.
20. Saving the research project $1,873.21 in mail mode costs.
21. Previous research by West (2014) suggests that legislators will sometimes defer "public service" tasks such as responding to surveys until after their legislative session ends. It was important to leave the Internet survey open to allow legislators this option.
22. This number assumes that the 183 legislators with returned letters did not participate in the online survey. Because of anonymity, this assumption cannot be verified.
23. The most recent Azavea and Cicero whitepaper available.
24. Currently Maine, California, Colorado, Arkansas, Michigan, Florida, Ohio, South Dakota, Montana, Arizona, Missouri, Oklahoma, Nebraska, Louisiana, and Nevada.
25. Meaning that they have no grievance or other rights under the Arizona state personnel system. These rights are typically afforded all other state employees.

Chapter 2

Mature Communication Technologies

OVERVIEW

While Internet-enabled communication technologies tend to dominate the political communication landscape these days, it is fair to suggest that the Internet will not displace mature communication technologies as the communication venue of choice for legislators anytime soon. Both the qualitative and quantitative data analyzed in this chapter tell the same story; meeting face-to-face with a constituent, being able to look into their eyes, read their emotions, and understand their body language are all important to legislators—much more important it turns out, than any form of Internet-enabled communication technology analyzed in chapter 4. There is a good reason for this; actually, there are tens, hundreds, thousands, or millions (depending on your perspective toward evolution) of reasons why face-to-face communications are so important. Simply put, humans are evolutionarily wired to prefer more mature forms of communication. While the Internet holds great promise for precipitating change in political communications and policymaking, the data in this chapter suggests that for now anyway, Internet-enabled communication technologies (and even more so communication technologies) are losing the evolutionary war for the hearts and minds of legislators.

Mature communication technologies include face-to-face meetings,[1] hard-copy letters and documents, and telephone calls. In general, mature communication technologies are communication technologies that have existed significantly longer than Internet-enabled communication technologies, and in general, mature communications are private in nature. With the exception of the telephone, all of the mature communication technologies have existed thousands of years (or longer in the case of verbal communications), and can be thought of as evolutionarily significant in the development of *Homo sapiens*. Before transitioning to a review of mature communication technologies, it is important to clarify a couple of definitional points.

With respect to written communications, it should be noted that hardcopy letters were treated as a form of mature communication technology, while e-mail is covered under Internet-enabled communication technology. While both forms of communication use the "written" (or typed) word, as indicated in the qualitative analyses results, legislators treat the two forms of written communication differently. The survey of state legislators used for the quantitative analyses in this chapter clearly differentiated hardcopy letters from electronic e-mail by specifying "non-electronic" hardcopy letters.

To clarify a second potential point of confusion, although the survey instrument did not specifically rule out Internet video conferencing from the face-to-face category of communicating, interviews with legislators suggest they share a common understanding that face-to-face meetings are meetings in which the participants are physically in the same room. With respect to telephone calls, the survey did not differentiate between landline phones, cell phones, and voice over IP (VIOP) communications. For the purpose of the analyses in this chapter, all three types of phone calls are acceptable even though the bandwidth of a phone call can vary significantly based on the type of transmission protocol used.

MATURE COMMUNICATIONS

From the perspective of human evolution, the communication technologies analyzed in this chapter vary greatly in the length of time they have existed. The three communication technologies: telephone invented in 1876 (Huurdeman, 2003) is the newest, writing is next with the earliest samples dated 3200 BC (Schmandt-Besserat, 1996), and face-to-face communications are as old as mankind itself. From an evolutionary perspective, it is likely that only face-to-face communications have existed long enough to impact modern human communication behaviors. For the vast majority (all?) of their existence, humans have communicated with each other face-to-face.

For the purposes of the analyses in this section, there is an assumption that the communications are private. In the interview process, legislators indicated that face-to-face, telephone, and hardcopy letters are assumed to be private (to the extent possible under Freedom of Information Act [FOIA] laws). Nowhere is this assumption clearer than in legislators' discussion of FOIA laws and how mature communication technology can circumvent these laws. This understanding is important because the assumption of semi-private or public communications using mature communication technology would likely lead to different regression results.

There are two important concepts that focus on the human preference for face-to-face/mature communication technologies: media richness and media naturalness. These concepts are presented in the next section.

Media Richness and Media Naturalness Theory

Media richness and media naturalness theories provide links that enable one to predict how a particular communication technology or range of technologies may impact human behavior. While there are many human behaviors related to communication technology that can be investigated, this section focuses on using media richness and naturalness theory to develop a hypothesis regarding the relative importance that legislators can be expected to assign to a range of communication technologies.

To state the concepts of media richness and media naturalness theory more simply, the richer the media the higher the bandwidth. Higher bandwidth is associated with media that is more natural (more face-to-face like). More natural, higher bandwidth media convey more information, which makes these media less ambiguous, more intellectually and physiologically stimulating, and decreases the effort (and time) required to understand the information being conveyed. Individuals should, all else equal, prefer less ambiguous information.

A basic understanding of both media richness theory and media naturalness theory requires a brief overview of the topic of social presence. Social presence is an important concept in media richness theory. According to Short et al. (1976), various communication technologies differ in their ability to communicate both quantity and type of information in a fixed timeframe (equivalent to the natural science concept of bandwidth). Short suggested that higher bandwidth communication technologies were associated with increased social presence while lower bandwidth media types were associated with lower social presence. According to Burke (K. Burke and Chidambaram, 1999) bandwidth is defined as the "range of [verbal and non-verbal] cues transmitted by the [communication] medium; a higher bandwidth medium transmits more types of cues than one with less bandwidth" (559). Social presence is defined as "the ability of learners to project themselves socially and affectively into a community of inquiry" (Rourke et al., 1999, 1). In effect, social presence is the extent to which a learner feels the presence of an individual with whom they are interacting. According to Rourke et al., social presence as a concept has its roots in Wiener and Mehrabian's (1968) concept of *immediacy*, defined as "those communication behaviors that enhance closeness to and nonverbal interaction with another" (203).

The essence of media richness theory is that different communication technologies vary in "richness," which is defined by Daft and Lengel (1986) as "the ability of information to change understanding within a time interval" (560). Daft and Lengel go on to state that communication technologies that require a long time on the reader's part to understand are less rich, while communication technologies that convey information quickly are richer. Time is an important factor both in media richness theory and in determining why a particular communication technology may be more or less important to a legislator, but why might time be important to legislators?

In *Information Sources in State Legislative Decision Making*, Mooney (1991) references work by March and Simon (1958) and Huber (1989) to justify the importance of time in legislative decision-making. Referencing legislative bounded rationality, Mooney suggests that because legislators have severe limitations on their time, they will search for the information they need to make a decision from the most readily available source. Like Mooney, Arnold (1992) lists a legislator's time (and that of their staff) as: "two of their scarcest resources" (36–37). Mooney goes on to suggest that once legislators obtain the information they need, they will stop searching. In Simon's terminology, legislators who acted thusly would be "satisficing" (1957, 119).

Associated with the shortage of time as a motivating factor for legislative information selection, Bradley (1980) in his research *Motivations in Legislative Information Use* found that legislators are "strongly motivated" (399) to use information sources that are both accessible and convenient. According to Bradley, in the legislators polled (n=36), the most important aspect of information is accessibility (72 percent), while convenience and understandability were tied as the second most important attributes of information. The link between legislators having limited time and the importance of accessibility and convenience of information is clear—logic would dictate that accessible, convenient information *should* be important for legislators who have limited time to address all of the tasks they face.

To summarize the hypothesis thus far: First, media richness theory suggests that the richer the media, the more information it can transmit over a fixed period of time. Second, legislators are time-constrained and value (read: find more important) information that is clear and concise, and can be gathered quickly. Third, and derived from the two previous relationships, legislators can be expected to find richer communication technology more important than leaner communication technology.

Burke (1999) outlines media richness theory which suggests that communication technology with inherently limited cue-carrying capacity will be less effective on ambiguous tasks than on simpler, pre-defined tasks (560). Media richness theory suggests that ambiguous information requires more bandwidth to be understood, while simpler information requires less bandwidth.

The richer the media, the more social presence that is communicated, and the higher the bandwidth of the communication technology needed to communicate the information. Empirical attempts to test the media richness hypothesis have resulted in mixed results, with some studies finding support for the theory and others finding little or no support (Kock, 2005).

Taking a different theoretical approach that suggests a relationship between communication technology and human behavior, Kock (2005), hypothesized that the "naturalness" of communication technology may directly impact human behavior. Kock defines naturalness as "degree of similarity to the face-to-face medium" (117) and suggests that the less natural a communication media is, the more effort humans must expend to understand the information that is being communicated. Specifically, Kock suggests that less natural communications increase cognitive effort, increase communication ambiguity, and decrease physiological arousal, "each of which may or may not lead to certain types of behavior or task outcomes" (125). To test his theory, Kock (2007) performed an experiment on 230 university students that compared the cognitive effort of face-to-face communications with a Web-based quasi-synchronous electronic medium similar to an interactive blog. Kock found that the web interface increased cognitive effort by 12 percent and communication ambiguity by 19 percent, and caused an increase in receiver effort by 19 percent over face-to-face communications.

Putting Kock's naturalness theory in terms of legislative behavior and communication technology, the less face-to-face like a communication technology is, the more cognitive effort a legislator must expend, the less physiologically aroused the legislator may be, and the more ambiguity there may be in the communication. If legislators are as time-constrained as many researchers suggest (Ellis, 2010; Harden, 2011; Kingdon, 1989; Oleszek, 2011), then communication technologies which require more cognitive effort are less physiologically arousing and more ambiguous, and can be reasonably expected to *decrease* the importance of that communication technology. If this is the case, then the more face-to-face like a communication technology is, the greater the importance that technology should be to a legislator. The question then becomes, "How might communication technologies be categorized by their 'naturalness'?" Recent literature drawing on media richness theory offers an answer to this question.

Mergel (2012), investigating social media adoption at the US federal level, builds on media richness theory to define a connection between the richness of interaction for various communication technologies. Mergel notes some of the advantages and disadvantages associated with various communication technologies in use by public sector entities and uses the term "informal interactions" to describe how rich or face-to-face like

Table 2.1 Communication Media Richness of Interaction

Communication Media	*Richness*	*Advantages*	*Disadvantages*
Formal Report	Low	Provides Records, Premeditated, Easily Disseminated	Impersonal, One-way, Time lag in feedback
Memos, Letters E-mail, IM, Web Phone, VOIP Social Media	↕	↕	↕
Face-to-Face	High	Personal, Two-way, Reflexive Feedback Cycles	No record, Spontaneous, Dissemination Difficult

Self-Generated by author using Microsoft Word.

certain communication technologies are. Mergel's ranking of the richness of various communication technologies provides a convenient platform upon which the importance (from a legislator perspective) of these communication technologies may be derived. Table 2.1 presents the relative media richness (from low to high) of some of the communication technologies investigated in this study. Table 2.1 can be compared to table 5.1, which highlights some of the primary and secondary characteristics such as duplex and bandwidth for the specific communication technologies examined in this book as an aid to predicting the importance legislators assign to various communication technologies.

QUALITATIVE RESEARCH

In addition to presenting the results of quantitative analyses, chapters 2 and 4 also present the results of recorded and professionally transcribed interviews with Arizona state legislators, legislator assistants, and Arizona state information technology (IT) staff that support legislators and legislator assistants. These semi-structured interviews explore the relationships between these three groups and communication technology in an effort to understand how communication technology influences behaviors that impact the policy process. These interviews help uncover some of the complex relationships that exist between legislators and communication technologies. Qualitative research is sometimes called "thick" research (Miles, Huberman, and Saldaña, 2014; Johnny Saldaña, 2015) because of its ability to reveal complex relationships, especially when it focuses on "naturally occurring" ordinary events in local contexts as the research in this chapter does. Simply put, the research in the next three chapters explores legislator use of communication technology by allowing legislators, legislator assistants, and IT support staff to use their own words to explain their relationships with communication technology.

Interviews presented in chapters 2 and 4 were completed several years prior to the quantitative research discussed in the other chapters in this book, and were conducted only with Arizona legislators[2]. The qualitative information presented in these chapters is meant to showcase the complexity of the relationships between legislators, legislator assistants, and communication technology by allowing legislators, legislator assistants, and IT staff at the Arizona capitol complex to explain how they use and value various communication technologies. Discussion begins with a brief overview of the qualitative methodology used to gather the data presented in chapters 2 and 4.

Qualitative Methodology

Nonprobability qualitative quota sampling was used to obtain a quasi-representative sample of legislators and legislator assistants with demographic characteristics of interest in this research. Nine legislators and their paired (nine) legislator assistants were interviewed, along with three staff from the IT department. An anonymized description of each of the interviewees is contained in appendix A.

Semi-structured interviews were utilized for these face-to-face interviews. Semi-structured interviews strike a balance between allowing a participant to choose the topic of discussion (an unstructured format) and a quantitative or closed-ended format in which the participant has no control over the topic discussed. For the purposes of understanding the impact of communication technology on the behaviors, roles, responsibilities, and understanding of constituents of the staff of Arizona House and Senate legislators, it was important to control and guide the topic, as well as to allow the interviewee the space needed to feel comfortable in expounding in areas that the researcher did not address, an essential part of exploratory research. In short, in exploratory research, a researcher may not be aware of all of the questions he/she needs to ask, and a participant must have the freedom to extend the research in unanticipated directions.

The interview instrument was broken into three sections comprised of non-directional open-ended questions. Section one of the survey is comprised of an introduction and recording authorization. Section two, making up the main body of the interview, consists of fifteen interview questions spread across seven response categories. These response categories include communication technology frequency of use and preferences (three questions), the perceived risks and benefits of communication technology (two questions), behavior (two questions), roles and responsibilities (one question), new communication technologies (one question), constituents (four questions), and communication strategy (one question with five subparts). The interview instrument concludes with an open-ended exploratory question, which asks the following: As an insider, and in your opinion, do you think that there are other important aspects of the relationship between legislators and communication technology that I have not touched upon in this interview? [If so] What are they? This final question is specifically designed to prompt the interviewee to reveal aspects of the relationships between legislators and communication technologies that were not covered in the previous sections. Section three contains closed-ended demographic questions.

Reliability and Validity

Intercoder reliability testing occurred on 11 percent of the transcribed legislator interviews and utilized Krippendorff's alpha (α) calculation. Krippendorff's alpha was calculated (118 codes, 2 coders, matrix format) to be .889, indicating an 88.9 percent intercoder agreement. These results exceed the 85 percent recommended by Miles et al. (2014, 85).

First pass coding utilized provisional coding (literature review based code), hypothesis coding (researcher hunches), and holistic coding (broad topic areas arising from the data). The second pass coding followed a focused approach, distilling first pass coding into themes based on conceptual themes related to communication technology relationships and behaviors. For first pass codes, a quasi In Vivo approach was utilized. This approach summarized coded sections into short sentences that could easily be understood in a network diagram and was designed to capture the nuances associated with communication technology frequency of use and behaviors. Quasi In Vivo second pass codes were typically summarized with one or two word codes that were then "connected" to first pass codes to determine theoretical relationships based on network link densities. In effect, clusters of codes with similar theoretical themes appear as node clusters in network diagrams.

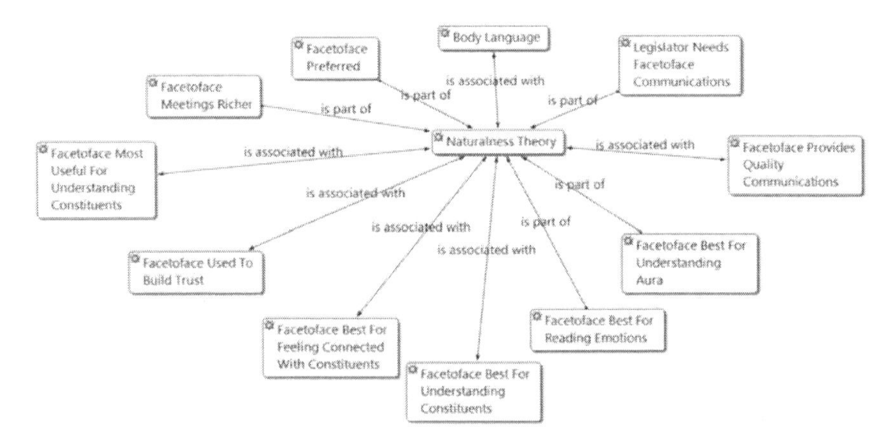

Figure 2.1 Face-to-face naturalness theory network diagram.
Self-generated by author using AtlasTi Qualitative Analyses Program.

Figure 2.1 contains an example of a network diagram (in this case, naturalness theory associated with e-mail communications) which clusters first and second pass codes into analytical themes. Several dozen network diagrams were generated as part of the qualitative analyses; however, the vast majority was too large and too complex to print in a readable format in this book. However, Rowman and Littlefield have added a full set of network diagrams to their web page for this book.

Among other topics, interviews focused on frequency of use and importance of various communication technologies. The topic of importance can be problematic because the concept of importance is both subjective and somewhat ambiguous. Importance was defined for legislators as, "Importance is related to the likelihood that you will respond favorably to a request received from a peer (or constituent) via one of the communication technologies." With respect to qualitative research, legislators were allowed to define importance however they wished because their elaboration on importance added significant value to the study. Table 2.2 outlines the importance that legislators associate with various communication technologies and their legislative duties as defined in network diagrams for all communication technologies examined during interviews. Table 2.1 and 2.2 can be compared to table 5.1, the theoretical importance of the various communication technologies examined in this book.

This concludes the qualitative discussion necessary to aid in the understanding of the qualitative data presented in chapters 2 and 4. With this step complete, focus now turns to a presentation of the legislator interviews associated with mature communication technologies.

Table 2.2 Communication Technology Importance and Legislative Duties

Survey Importance Rank	*Communication Technology*	*Interview Importance Associated With*
1	Face-to-face Communications	Trust Building, Influence, Coalition Building, Reading Body Language, Privacy, Job Satisfaction, Understanding Emotions, Feeling Connected, Understanding Individual, Passing Legislation, Significant Topics, Circumventing FOIA Laws, Impact On Constituents, Mobilization, Constituent Feedback
2	e-mail	Efficiency, Convenience, Taking Care of Routine Legislative Business, Mass Communication, Disengagement, Constituent Impact, Time Savings, Coalition Building, Campaigning, Mobilization, Constituent Feedback (Except Boilerplate e-mail)
3	Telephone	Hearing Emotion, Efficiency, Multitasking, Time Savings, Feeling Connected, Addressing Constituent Issues, Circumventing FOIA Laws, Significant Topics, Reaching Older Constituents, Impact On Constituents, Mobilization, Constituent Feedback
4	Letters	Constituent Recognition, Significant Documents, Maintaining Contact With Low Technology Use Constituents
5	Facebook™	Campaigning, Releasing (Pushing) Information Quickly, Mass Communication, Running For Higher Office, Chamber Leadership
6	Web Page	Campaigning, Obtaining Information (Research)
7	Twitter™	Campaigning, Releasing (Pushing) Information Quickly, Mass Communication, Running For Higher Office, Chamber Leadership
8	Blog	Campaigning, Releasing (Pushing) Information Quickly, Constituent Feedback

Self-Generated by Author using Microsoft Word.

LEGISLATOR INTERVIEWS

Face-to-face Communications

Face-to-face network diagrams cluster codes around primary themes suggested in second pass coding based on first pass grounded codes. Second pass themes for face-to-face communications include importance, trust, constituents, priorities, legislator assistants, risks, lobbyists, communication technology frequency of use, legislators, and naturalness theory. Face-to-face communication behavior is grounded in 2,837 quotes and 83 individual codes.

Importance of Face-to-face Communications

Legislators frequently mentioned that face-to-face meetings were the most important communication technology they used. Legislator J sums up her feelings on face-to-face communications: "I think in the legislative process still person to person communication whether that be via telephone or in person, face-to-face is the most important." Only one legislator, legislator K, indicated that another communication technology was more important than face-to-face:

> I would say e-mail would probably be at the top of the list, now. Fifteen years ago I never would have thought I would say that. But, yeah, that would be a high priority and important to me. The face-to-face that I do is important, but like I say, I really, really limit it. Just because of my own personal communication style. But it's important in the sense that when I say, you know, Lobbyist A wants to talk about this particular bill. I may be really anxious to talk to Lobbyist A about that bill because they may be able to share something that I just don't otherwise know or know even how to figure out.

Face-to-face network diagrams suggest a number of reasons why face-to-face meetings are important, but one of the most interesting is that the more important a topic is to a legislator, the more important face-to-face communications become. Four legislators tied the importance of the topic being discussed to the importance of face-to-face meetings. Here is legislator B with a typical response to why face-to-face meetings are important:

> Body language, expression, emotional reaction, I function a lot off emotional, they call it emotional intelligence, they call it emotional cues, so if I am wanting to engage somebody in something that's important to me, I will make a point of having a face-to-face communication, if it is simply a matter of disseminating information, or gathering information, e-mail is fine.

Importantly, legislator B includes some of the reasons why face-to-face meetings are important to her, and being able to read body language tops the list. One of the reasons that body language is so important is because they believe it builds trust. Legislators find that the ability to read body language is an important factor when building trust with another individual, because it gives them a more complete picture of what the other individual is feeling.

TRUST

Legislator L on the topic of trust and face-to-face meetings:

> Well, I think that comes in part of the communication well before you ever get on the floor and you start doing that—, that face-to-face communication and—, and people that you've built trust with. And so they've been in your office before the bill ever goes to—, to the floor and you've discussed and talked about those kind of things. And so you know who you're getting the information from and so you can—, you can rest pretty assured that they're, you know, telling you the—, the—, the correct information that—, that will assist you in—, in what you need.

In the passage at the end of legislator L's response, he struggles to say the word "truth" and instead, stammers out "the correction information" rather than simply using the word "truth." The discussion continues with why face-to-face meetings are important:

> Respondent: Well, I think—, you know, I'm a—, I'm a face-to-face type guy. A lot of people, you know, they talk on the phone and they can-, you know, that. But I want to—, I want to be able to look you in the eye and I want to be able to, you know, see your reactions as—, as I ask questions and as we talk about certain things. And—, and I want you to be able to respond back to me when I ask questions that—, that I think are pertinent to the legislation that you may be for or against in that. So I try to build that personal relationship before I ever to get the floor and start, you know, discussing or passing or voting on any legislation at all.

In effect, legislators use face-to-face meetings to determine the trustworthiness of the individual they are meeting with. The reason face-to-face meetings are important for building trust is because legislators believe it is more difficult to hide true feelings when body language and/or vocal clues are available.

Constituents

If there is a single summarization of the relationships between legislators, face-to-face meetings, and constituents, it is that such meetings are rare. While legislators meet routinely with large groups of constituents, they do not generally sit down one-on-one with them. The face-to-face network diagrams offer clues to some of the reasons why the meetings are not convenient, face-to-face meetings intimidate constituents, and face-to-face meetings are not necessary for the majority of the reasons why a constituent may contact a legislator. Four of the nine legislators indicated that face-to-face meetings were rare, including legislator K: "Yeah. I probably have not sat face-to-face in eleven years with but a handful of constituents."

Priorities

Legislators prioritize group and one-on-one meetings with constituents as a top priority, frequently indicating that being seen by large numbers of their constituents as being a high priority for them. Here is legislator G prioritizing communication technologies:

> Yeah, I would say in ranking of order, face-to-face, one-on-one, a group face-to-face is probably second, then I would say probably a phone call, then an e-mail, in terms of priority. And then probably Facebook™ and then probably website would be last, so the least interactive.

All but one legislator listed face-to-face meetings as their top priority for communication format, whether the meetings were one-on-one or group meetings, or whether they were with lobbyists or constituents. Only one legislator, legislator K, indicated that he limits face-to-face meetings.

Legislator Assistants

Not surprisingly, the interactions between legislator assistants and face-to-face meetings with legislators are largely related to the task of scheduling meetings for the legislator. One interesting exception to this is that many legislator assistants screen face-to-face meetings for the legislator they work for. Here is legislator D discussing how her legislator assistant screens face-to-face meetings:

> They have those, you know. Call whoever and maybe it's being sent by your opponent to call in. So, you know, she's good at trying to find the source of the calls. And whether people are asking for an appointment and whether or not I

should meet with them. You know, all those things I think [name deleted] is a great help in deciphering where the source is, is it going to be a benefit to them to meet with them, or not?

In the quote above, legislator D raises an interesting point that political opponents will call in to obtain information that will give them a political advantage. The concept of how communication technology is used to gather information to be used to gain political advantage may be an interesting topic for future research.

Risks

Of all the communication technologies examined, legislators listed face-to-face communications as the least risky; only one legislator associated risks with face-to-face communications. Legislator H: "One of the risks is e-mail and face-to-face is the hearsay that comes out of being face-to-face; you know, you tell somebody something and you never know if it stays where it's supposed to stay, and that's how rumors get started." The legislator mentions two risks but appears to conflate them. Hearsay is the risk associated with the inability to adequately verify the information being communicated. The second risk that legislator H mentioned is the inability to control distribution of the information being communicated, which was a significant risk listed for e-mail.

Lobbyists

Results of the interviews suggest that face-to-face meetings occur primarily with other legislators and lobbyists. Network diagrams highlight two nodes that are most interesting from a communication technology perspective. First is that many legislators prefer to meet with lobbyists face-to-face because they need the additional information contained in a legislator's body language. Here is legislator I on the topic:

> I do. And I've noticed with lobbyists, what lobbyists do, and they understand this, is that they do not do anything. They are very careful to just be as neutral and vanilla and not show any emotion one way or another whatsoever. So, I think they are, you know, I mean, it depends, they may, you know, if they really want something they may show emotion. But I would say that lobbyists are very careful not to ever show a negative. If they disagree with you, unless it's the exact issue they're talking about, they're not going to let you know that. They want to be everything to everybody.

Legislator I on the same subject:

> I get more . . . yeah, I get more feedback, I get more emotional feedback, I read body language a little better and know how they're . . . you know, like when I'm talking to a lobbyist, face-to-face on an issue is so much better than an e-mail, because I can figure out a little bit about where they're coming from, whether they're in favor of it or whether against it. You know what I'm saying?

Legislator I suggests that lobbyists intentionally try to hide their body language in order to obfuscate their true feelings on a topic. Legislator I reinforces the concept that reading a lobbyists' body language reveals their feelings on an issue, and provides a comparison between face-to-face and e-mail, noting that face-to-face is better than e-mail for uncovering a lobbyist's feelings toward an issue. Whether a majority of face-to-face meetings are actually with lobbyists or with other legislators appears to be a function of the legislator's preference, and is briefly discussed in the next section on legislator behavior.

Legislator Behavior

Some of the face-to-face network diagrams have been discussed previously due to their linkages to other network node concepts. Of the behaviors that remain to be explored, there are two that are most interesting: age relationships and position flexibility. Theory suggests that older legislators may use newer communication technologies less frequently. The corollary to this finding is that older legislators will use mature communication technologies more frequently: and this appears to be the case. Five interviews indicated that older legislators preferred face-to-face communications because of their age, using terms like "old school" to describe themselves and effectively tying their behavior to their age.

A second unexplored topic regarding legislator behavior and face-to-face meetings involves the "position flexibility" that face-to-face communications offer a legislator. Legislator L explains the following:

> Because as you—, as you—, in my case anyway, when I start to have a conversation and I ask a question then I can—, I can start to see which direction. And so then I can either move in that direction or this direction, depending on the responses and so on that I get. And I—, and I've always been a—, a—, a face-to-face guy. So that's the one I use the most.

Here we see that legislator L adapts his dialog based on feedback he receives on the position taken by the person he is meeting with. This positional flexibility is related to full-duplex communications (real-time interactivity) and is one of the benefits of immediate feedback and feedback based on body language, a key component of naturalness theory.

Naturalness Theory

The links between naturalness theory and the behaviors that drive a legislator's use of face-to-face meetings are among the most interesting of this study. Figure 2.2 contains the network diagram reflecting the relationships surrounding face-to-face communications and naturalness theory presented earlier in this chapter.

The discussion of face-to-face communications in the context of naturalness theory begins with a legislator's need for face-to-face communications. Legislator B:

> I still have a tremendous need to have face-to-face communications and interactions with people. And part of that is that I am a very intuitive person, I need to be able to read someone's reaction to my ideas and to my communication.

Naturalness theory suggests that humans have an evolutionarily driven need for face-to-face communications because they offer significant physiological arousal (Kock, 2005, 123). Legislator I, when asked what communication technologies makes him feel more closely connected with constituents, said this:

> Well, I think face-to-face, again, is going to be the best. But and right after that is probably speaking over the phone, and then Facebook™.

Legislator I lists his "feeling connected" preferences from most fulfilling to least fulfilling, in the exact order that naturalness theory would predict—from the most physiologically arousing to the least physiologically arousing.

Legislator H, unsure of how to communicate the essence of what she picks up from the person she meets with face-to-face, calls it aura:

> Well, face-to-face always gives me the benefit of knowing . . . reading their personalities or aura or whatever it is about them; how they react to my statements, much better than e-mail. On the phone you can do it a little bit, you know, but face-to-face always seems to work better as far as being able to read their reactions to an issue.

Although the definition of the term "aura" involves an invisible emanation from a living creature, it is reasonable to suggest that legislator H is speaking about non-verbal human emotions. Three legislators relate the ability to understand another individual's emotional state with face-to-face meetings. Here is more of the interview with legislator H:

Okay. Tell me why you prefer face-to-face.

> Because I feel like the connection is better and that I can read faces and reactions to different issues; I just think it's better than an e-mail. I get more . . . yeah, I get more feedback, I get more emotional feedback, I read body language a little better and know how they're . . . you know, like when I'm talking to a lobbyist, face-to-face on an issue is so much better than an e-mail, because I can figure out a little bit about where they're coming from, whether they're in favor of it or whether against it. You know what I'm saying?

Legislator I continues the theme of understanding emotions through face-to-face meetings:

> Yeah, face-to-face is-, face-to-face is always going to be the best way to completely understand somebody, I think. I think body language is, I think, yeah, I do think so. And emotions and those kinds of things.

The link between naturalness theory and human emotions is clear. Humans, evolutionarily programmed to respond to the emotions of those around them, find face-to-face interactions provide the best possible link for understanding emotions (Kock, 2005). In addition, Kock notes that humans can expect to find face-to-face interactions more emotionally fulfilling than computer-mediated communications. Legislator E, searching for a term to describe face-to-face meetings, calls them "quality": Well, I think for quality communications I prefer face-to-face, but for speed efficiency and quantity, e-mail is just a godsend.

Legislator E captures one of the common relationships between legislators and e-mail; e-mail is useful for speed, efficiency, and quantity.

The above interview quotes highlight the close relationship between naturalness theory and a legislator's use of face-to-face meetings. In addition, legislators, in the process of identifying key attributes of face-to-face meetings, note the relationships between face-to-face meetings and other communication technologies. These relationships are predicted by naturalness theory, and are consistent with the results obtained during the quantitative survey.

Telephone Communications

Just as with the face-to-face network diagram, the telephone network diagram clusters codes around primary themes suggested in second pass coding based on first pass grounded codes. Second pass themes for face-to-face communications include phone banks, communication technology frequency of use, importance, phone versus e-mail, naturalness, constituents, age, legislator assistants, and legislator behaviors. Phone behavior is grounded in 1,689 quotes and 123 individual codes.

Phone Banks

Discussion on phone banks was limited to two legislator assistants, both of which indicated that they had little patience for them. Essentially, phone bank communications occur when grassroots organizations such as the American Association for Retired Persons (AARP) pay a phone bank to phone their members and ask them to communicate with their legislators regarding a specific topic, suggesting what the member should say. The phone bank then connects the member to their legislator via a phone transfer. Here is assistant B on the topic:

> —nonstop. Now, there are different groups who will—, and I'm very much opposed to this. I really don't like this, for a perfect example. AARP will call somebody's home—, somebody—, one of their AARP members and ask them if they support the governor's Medicaid expansion. And I don't know the whole shpeal that they give them—but whatever it is—, and the person will say, "Well, yes." not—, not realizing what's happening. And then the next thing they know they're transferred to my phone. I answer the phone and these are senior citizens, mind you. And they're like, "How did I get you? I didn't call you." And, you know, they start thinking government conspiracy.

Assistant I:

> Yes. We had one recently that was very interesting the way they did it. They contacted the person first and told them you—to call their representative or their Senator and tell them you support blah, blah, blah. Or let's say support Medicaid expansion because that was the one that did it. So—and press 1. And when they press 1, we get the ring and answer it and say Representative So-and-so's office. And they say, "Huh?"

Both legislator assistants note that constituents are confused by the AARP phone bank methods, and both note that constituents take policy positions against the policy position requested by AARP. Legislator assistants interviewed see phone banks as the telephone version of bulk e-mail.

Frequency of Use

Both legislator assistants and legislators note that phone lines are much busier when the legislature is in session than when it is out of session. Here is an interesting exchange with legislator E, who is responding to a frequency of use question:

> Okay. All right. . . . I guess along the lines of the communication you use on a daily basis, what would you say, which communication technology do you use the most?

E-mail.
E-mail? How about next . . . what's the next most?
Well, let me take that back; face-to-face is first, e-mail is second and then probably telephone.

Legislator E responds with e-mail as the most frequent (which agrees with both the quantitative survey and qualitative word count data) but, for some reason, changes his mind and indicates that face-to-face is the most frequently used communication technology.

Importance

In general, the importance of the phone was tied to specific tasks, such as emergency communications or filling in information gaps quickly. Legislator L listed the phone and face-to-face meetings as most useful for understanding constituents. Assistant A and legislator I both indicated that phone and e-mail communications were the most important communication technologies. The topic of phone versus e-mail came up frequently, even though there was no specific question asking interviewees to compare and contrast these communication technologies.

Phone versus E-mail

Several legislator assistants noted that over time, e-mail has (in their opinion) decreased the frequency of telephone calls and faxes. Assistant I, assistant B, and assistant D all noted this effect. Here is assistant D:

Faxes. You don't get many faxes.
How about phone calls? Tell me about phone calls.
Used to be—, used to be the phones would ring off the hook but with e-mail taking up more—, becoming more and more convenient for people-
Mm-hmm.
—we get more e-mails now. But I still get a lot of phone calls every day. I don't have a number that I could give you because it—, again it depends on the issue.

Several interviewees indicated that the phone is more personal than e-mail; which is a naturalness concept. Here is assistant F on the subject:

Me personally, I like e-mail, I like computers, but it's just very impersonal—
Mm-hmm.
—for me. I prefer to pick up the phone—

Naturalness

There are several concepts that link phone use to naturalness theory. Interviewees indicate that the phone is more personal than e-mail (as naturalness theory would predict). Most legislators noted that face-to-face communications are best for understanding people, but one legislator assistant indicated that the phone was best, but quickly followed up with face-to-face meetings. Figure 2.3 highlights the network diagram associated with telephone naturalness theory. Here is assistant D on the topic of telephone communications:

> Well, the phone is always the best because—, or if they actually come down here in person. And we do have that at times.

Here is legislator J on the topic:

> Important as far as influencing people and getting the job done, face-to-face communications, directly to a legislature or constituent whether that be over the face-to-face in person is the best, next would be over the phone, next is e-mail, and then I just view Facebook™ and Twitter™ as a way for me to get my message out. And the newsletter and the guest columns and all that is a way for me to get my information out, but also a mechanism for them to respond, I do like feedback from constituents because it is important to me.

Importantly, legislators indicated that they avoid contact with constituents who disagree with their political ideology and/or policy position only for certain communication technologies. Most legislators indicated that they did not have the time to engage with constituents who disagreed with their policy position and/or ideology via e-mail, more legislators indicated that they would engage constituents on such topics over the phone, and *every* legislator interviewed on the topic indicated they welcomed face-to-face engagements with constituents who disagreed with them on ideological or policy issues. Legislator challenge avoidance is a function of communication technology. The less natural the communication technology, the more likely a legislator is to avoid challenges from constituents via that communication technology.

In several codes, the phone is shown as listed second best or second most important. In all seven quotes associated with these codes, face-to-face communications were listed first. Naturalness theory predicts this relationship, as phone conversations are less "face-to-face like" than face-to-face meetings.

Legislator J:

> Okay, so e-mail is one of the primary ways you communicate?

> I would say e-mail is definitely, I think in the legislative process still person to person communication whether that be via telephone or in person, face-to-face is the most important.

Legislators note another key feature that makes telephone communications more natural; it communicates information such as emotions and "Aura." Here is assistant D:

> And I'll say, "I have your e-mail but I need to ask you a few questions." You know, I'll give them a reason why I'm calling them and stuff. And so then I can hear their voice. I can know if it's urgent, if it's not so urgent. You know.

And legislator H:

> Well, face-to-face always gives me the benefit of knowing . . . reading their personalities or aura or whatever it is about them; how they react to my statements, much better than e-mail. On the phone you can do it a little bit, you know, but face-to-face always seems to work better as far as being able to read their reactions to an issue.

Legislator Assistants

Answering a legislator's office phone is one of the single most important aspects of a legislator assistant's job. Although legislators are not typically in their offices during the out-of-session time periods, legislator assistants are required to be in their office and answer the legislator's phone, year-round. The large number of nodes in the telephone network diagram[3] is an indication of the importance of the phone to a legislator assistant's job. Several of the codes indicate the phone's priority: legislator assistants must answer the phone, phone most important for legislator assistant, phones transferred during lunch to ensure a live response. No legislators indicated that they answer their own office phone, and many indicated that their assistants screen (or filter) their phone calls. Here is legislator D with a typical response:

> Well, I'm learning, you know. It takes a while to learn who is the kind of person that picks up the phone rather than does the e-mail thing. Most people do e-mail, now. So, when somebody does pick up the phone, [name deleted] answers my phones and she helps screen them so that I know, you know, kind of the importance of—, and how to respond.

Legislator Behaviors

A legislator's phone is one of the key communication links with their legislator assistants. In important or urgent situations, legislator assistants will either call or text message their legislator to communicate the message. Some legislators publish their personal phone number on their campaign web pages and provide them to lobbyists, while some legislators do not give their phone numbers out to anyone other than family and other legislators, refusing to even allow trusted lobbyists to call them.

As with the phone constituent network diagram, the phone legislator network diagram has a quote to node ratio close to one. In effect, there is roughly one quote for every node.

Risks

As with face-to-face meetings, legislators mention few risks associated with telephone usage, and no risks associated with cell phone usage. Only one legislator indicated that phone conversations carry risk. Here is legislator B on the topic of risk:

> Well, everything that has an advantage has a disadvantage. If you put information out, you have no control over where it goes. You have to be very comfortable with that information. And all of the communications we do in the legislative setting is access to public examination. Request for information from the media.

Legislator B alludes to the fact that the media (or anyone) can request state legislator official phone logs under FOIA laws. Although US Presidential candidate Hillary Clinton was famously investigated for using a private e-mail server, interviews with legislators suggest that this is a common behavior and is meant to circumvent FOIA laws. In an interesting twist, when discussing risks, legislator B indicates a reason why phones are less risky than other written forms of communication:

> Yeah, well I'm cognizant of the fact that anything I write on a legislative e-mail account can be in the newspaper the next day. And very often I'll just tell someone to call me, just to avoid the so-called paper trail. Not that I'm doing anything illegal, but. . . . Yeah, but I mean, if I'm taking a position against a bill that might be generally popular or rather, if I'm taking a position for a bill that's unpopular, even though it's the right thing to do, I don't mind taking the hit if I have to vote on it. But if the bill winds up not getting out of committee and not going to the floor, then I don't want to take the hit, you know.

The qualitative section of this chapter has provided insights into the complexities surrounding legislator use of mature communication technology. The next section focuses on developing an understanding of the implications of communication technology on the policy process through quantitative rather than qualitative analyses.

QUANTITATIVE ANALYSES

In this section, constituent and peer communication inferential statistics associated with mature communication technology frequency of use and importance are used to explore legislator communication behaviors. Each of the following sections on peer and constituent communications begin with descriptive statistics on the frequency of use and importance of mature communications and conclude with predictive linear and nonlinear models which examine the relationships between mature communications and demographic, political, and institutional (DPI) variables. Analyses begin with how legislators communicate with each other using mature communication technologies. Importantly, only a small sampling of the results is explained for each Internet-enabled communication technology in the following sections.

DEMOGRAPHIC, POLITICAL, AND INSTITUTIONAL EFFECTS

Peer Communications

Legislators were asked to rate the frequency of use of and the importance they assigned to mature communications with their peers (other legislators). Importance was defined for legislators as "Importance is related to the likelihood that you will respond favorably to a request received from a peer (or constituent) via one of the communication technologies as shown in table 2.3. Frequency of use categories were "Do Not Use," "Use Annually," "Use Monthly," "Use Weekly," "Use Daily," and "Use Hourly." Importance categories were "Do not use," "Not Important," "Slightly Important," "Moderately Important," "Important," and "Very Important." Both frequencies of use and importance categories were coded 0 through 5, where 0 was do not use and 5 represented the highest frequency of use and importance. Table 2.3 highlights the frequency of use and importance of legislator use of mature communication technologies when communicating with their peers. Constituent communications will be covered in the section following peer communication frequency of use and importance.

Referencing table 2.3 when communicating with other legislators, legislators most frequently use the telephone, followed by face-to-face meetings and then written letters. Legislators find face-to-face meetings most important when communicating with their peers, followed by the telephone, and written letters. Legislators find face-to-face meetings most important, but they use the telephone more frequently when communicating with peers.

Table 2.3 Frequency of Use and Importance Peer Descriptive Statistics

Frequency of Use	*Face-to-face*	*Telephone*	*Hardcopy Letters*
Do Not Use	9	24	264
Use Annually	34	36	250
Use Monthly	303	190	528
Use Weekly	539	580	380
Use Daily	612	708	199
Use Hourly	153	116	16
Mean/Mode	3.315/4	3.37/4	2.03/2
σ	0.978	0.953	1.27
Importance			
Do Not Use	4	8	264
Not Important	7	23	250
Slightly Important	32	87	528
Moderately Important	99	222	380
Important	407	694	199
Very Important	1072	589	16
Mean/Mode	4.54/5	4.06/4	2.03/2
σ	0.978	0.953	1.27

Self-Generated by author using Microsoft Word.

Inferential analyses start with bivariate ordinal regression models to examine the relationships between frequency of use (dependent variable) and importance (dependent variable) for each form of mature communication technology when communicating with peers, and the various DPI variables that form the independent variables for each ordinal bivariate regression. These variables include age, years in office, chamber (Senate or House) gender, education, race, nine-category political party (a measure of conservativism from strongly progressive [coded as 0] and strongly conservative [coded as 8], gerrymandering measures, majority/minority party status, polarization [binary and ordinal forms], professional/citizen legislature [binary and ordinal forms], term limits, conflict scale, and Trustee/Delegate scale).

Mature Communication Technology Frequency of Use with Peers

Bivariate ordinal logistic regressions were run for the importance of mature communications when communicating with peers, and DPI variables. Bivariate regression results suggest the complexity surrounding communication technology frequency of use and DPI variables. Note that the statistically significant relationships vary as a function of the type of mature communication technology being used. The following sections discuss these results in detail.

Face-to-Face Communications

For face-to-face communication frequency of use with peers, age, years in office, being a Senator, gender (male), and gerrymandering are associated with increases in frequency of use, while race (both black and the other race category) is associated with decreases in the frequency of use of face-to-face meetings. The largest effect found was the difference between the use of face-to-face communications by blacks and whites, with blacks being 48.8 percent less likely to choose "Use Hourly" for face-to-face communications with their peers over all other frequency of use categories than whites, on average, all else equal. The smallest effect found was the comprehensive gerrymandering measure where a one unit increase in gerrymandering scores (less gerrymandering) was associated with a 0.4 percent increase in the likelihood that a legislator would choose "Use Hourly" for face-to-face communications overall other frequency of use categories, on average, all else equal. Four of the five gerrymandering measures (comprehensive, Polsby-Popper, Convex Hull, and Reock) had statistically significant relationships with the frequency of use of face-to-face meetings. All of the statistically significant gerrymandering relationships indicate that a *decrease* in gerrymandering is associated with an *increase* in face-to-face communications. This is an intuitive result: the more political party homogeneity in a district, the less necessary it becomes to communicate with peers, since a legislator's district can be thought of as "safer."

With respect to gender, some studies suggest that there are little differences in how male and female legislators communicate (Thomas and Welch, 1991), while other researchers suggest that male and female legislators communicate differently (Thomas and Wilcox, 2014). Both Thomas and Welch and Thomas and Wilcox suggest that the *content* of legislator communications varies (or not) as a function of gender, but neither examine gender-based communication technology differences. No research was located which examined such differences, either with respect to peer or constituent communications.

The age findings associated with this regression are consistent with expectations. Research suggests that older individuals are less likely to use Internet-enabled communication technologies and in a broader context, computers in general (Juznic, Blazic, Mercun, Plestenjak, and Majcenovic, 2006; Schleife, 2006; Thayer and Ray, 2006). If one assumes that legislators communicate with approximately the same frequency across all forms of communication technology, then a decrease in Internet-enabled communication technology use could very well lead to an increase in the use of more mature communication technologies.

No research could be located which examined the use of face-to-face communications as a function of race (blacks use face-to-face communications with peers less than whites), legislative chamber (Senators use face-to-face communications with peers more than Representatives), or years in office (the longer a legislator is in office, the more likely they are to use face-to-face communications, but this may be related to the correlation between age and years in office), so these findings cannot be verified using secondary data or existing theories.

Telephone

The telephone communication frequency of use findings suggest that years in office, being a Senator, political party (a measure of conservativism), polarization, being male, being black, the Reock gerrymandering scale, professional legislatures, and acting more like a Delegate are all associated with increases in the frequency of use of telephone calls. Being a minority party in both legislative chambers and in the governorship is associated with a decrease in the frequency of use of the telephone compared to their majority party peers. The largest effect size once again was race, with blacks being 91.7 percent more likely to choose "Use Hourly" over all other frequency of use categories for telephone frequency of use compared to whites, on average, all else equal. The smallest effect size was the Trustee/Delegate scale, where a one unit (1 percent) increase in the legislator's self-evaluation as a Delegate is associated with a 0.3 percent increase in the likelihood that a legislator would report "Use Hourly" over all other frequency of use categories for telephone communications, on average, all else equal.

One of the more interesting findings associated with the above bivariate regression is the correlation between race and telephone use. Specifically, while blacks were significantly less likely (by roughly half, 48 percent) to use face-to-face meetings to communicate with other legislators, they were much more likely (almost twice, 91.7 percent) to use the telephone to communicate with their peers.

As with face-to-face communications, no research was located which examined telephone communications as a function of legislator years in office (the longer a legislator is in office, the more likely they are to use the telephone to communicate with peers), chamber membership (Senators use the telephone to communicate with peers more than Representatives), gender (male legislators use the telephone to communicate with peers more than female legislators), conservativism (the more conservative a legislator, the more likely they are to use the phone to communicate with peers), gerrymandering, minority party status (legislators who are in the minority party in both chambers and the governorship communicate with their peers less than those in the majority).

One expected, and perhaps not as interesting as the previous results is the relationship between telephone use and professional legislature status.

Professional legislators are more likely to use the telephone to communicate with other legislators than their citizen (part time) legislator counterparts, although interestingly, this relationship was not found to be statistically significant for either face-to-face or hardcopy letter communications.

Hardcopy Letters

For hardcopy letter communications with peers, years in office, education, the Reock gerrymandering score, and being a minority party in one chamber are all associated with increases in the frequency of use of hardcopy letters, while only the Convex Hull gerrymandering measure indicates a decrease in the use of hardcopy letters. The largest effect size was one chamber minority status. Legislators who were in a legislature where they were the minority party in one chamber were 48.5 percent more likely to choose "Use Hourly" over all other frequency of use categories for hardcopy letters compared with their majority counterparts, on average, all else equal. The smallest effect size was the Convex Hull gerrymandering measure where a one unit increase in gerrymandering (less gerrymandering) is associated with a 1.2 percent decrease in the likelihood that the legislator would choose "Use Hourly" for hardcopy letter use over all other possible frequency of use selections, on average, all else equal.

With respect to education, only hardcopy letters were statistically significantly related to a legislator's education. The more educated a legislator, the more likely they were to use hardcopy letters to communicate with their peers. This section reviewed regressions with a single DV (frequency of use of mature communication technology to communicate with other legislators); in the next section, three multivariate regression models will be analyzed using the frequency of use of mature communication technology to communicate with other legislators (*pfreqmature*) as the dependent variable.

Peer Communication Technology Frequency of Use Models

Three models were created to determine the relationships between peer frequency of use of mature communication technology and DPI variables. These models are presented in order to highlight how the statistically significant relationships change with increasing model complexity. For example, in the basic model, age is statistically significant. However, in the intermediate and advanced models, the addition of other variables has removed the significance of the age variable and increased the coefficient and significance of the years in office variable. In effect, the collinearity between age and years in office was masked in the basic model. The use of basic, intermediate, and advanced models is a common social science standard, and has nothing to do with the fact that three mature communication technologies were analyzed. The basic model contains legislator demographic information, the intermediate model adds primary political and institutional variables, and the advanced model adds secondary institutional variables. Additionally, the results of these three models are not presented in a table, but rather, are discussed in the paragraphs following the models presented below:

Basic Model: *pfreqmature* = $\beta_0 + \beta_1(age) + \beta_2(race) + \beta_3(male) + \beta_4(education) + \mathcal{E}$.

Intermediate Model: *pfreqmature* = $\beta_0 + \beta_1(age) + \beta_2(race) + \beta_3(male) + \beta_4(education) + \beta_5(conservativism) + \beta_6(minority\ status) + \beta_7(years\ in\ office) + \beta_8(senate) + \mathcal{E}$.

Full Model: *pfreqmature* = $\beta_0 + \beta_1(age) + \beta_2(race) + \beta_3(male) + \beta_4(education) + \beta_5(conservativism) + \beta_6(minority\ status) + \beta_7(years\ in\ office) + \beta_8(senate) + \beta_9(gerrymandering) + \beta_{10}(professional\ legislature\ binary) + \beta_{11}(term\ limits) + \beta_{12}(conflict\ scale) + \beta_{13}(Trustee/Delegate\ scale) + \beta_{14}(polarization\ binary) + \mathcal{E}$.

The three models all use a composite dependent variable for mature communications that is the median of each of the three individual mature peer communication technology frequency of use variables.

The basic model is statistically significant for age (odds ratio = 1.001, $z \leq 0.01$), gender (odds ratio = 1.233, $z \leq 0.05$), and education (odds ratio = 1.039, $z \leq 0.05$). Interpreting a single variable in the basic model for clarity; a one-year increase in legislator age is associated with a 1 percent increase in the likelihood a legislator will select "Use Hourly" for the use of mature communications with peers over all other communication technology frequency of use categories, ceteris paribus: The older a legislator gets, the more they will use mature communication technologies.

The intermediate model is statistically significant for education (odds ratio = 1.040, $z \leq 0.01$), majority party status in one chamber (odds ratio = 1.559, $z \leq 0.01$), and years in office (odds ratio = 1.829, $z \leq 0.001$). Interpreting a single variable in the intermediate model for clarity, a one-year increase in the number of years a legislator has been in office is associated with an 82.9 percent increase in the likelihood a legislator will select "Use Hourly" for the use of mature communications with peers over all other communication technology frequency of use categories, ceteris paribus: The longer a legislator is legislator is in office, the more they will use mature communication technologies, even *after* controlling for age. This particular finding is important because it suggests that the increase in the use of mature communication technologies is both a function of age *and* years in office.

The full model is statistically significant for education (odds ratio = 1.044, $z \leq 0.05$), majority party status in one chamber (odds ratio = 1.430, $z \leq 0.05$), years in office (odds ratio = 1.030, $z \leq 0.001$), and being in a professional legislature (odds ratio = 1.428, $z \leq 0.01$). Interpreting a single variable in the full model for clarity, a legislator being a member of a professional legislature is associated with a 42.8 percent increase in the likelihood a legislator will select "Use Hourly" for the use of mature communications with peers over all other communication technology frequency of use categories, ceteris paribus.

Full Regression Models, Mature Communication Technology Frequency of Use with Peers

The previous sections analyzed and discussed mature communication technology relationships as a result of bivariate and multivariate regressions. In this section, the frequency of use and importance of each mature communication technology is analyzed using full regression models of the form:

$$\textit{Peer Mature Technology Frequency of use} = \beta_0 + \beta_1(\textit{age}) + \beta_2(\textit{race}) + \beta_3(\textit{male}) + \beta_4(\textit{education}) + \beta_5(\textit{conservativism}) + \beta_6(\textit{minority status}) + \beta_7(\textit{years in office}) + \beta_8(\textit{senate}) + \beta_9(\textit{gerrymandering}) + \beta_{10}(\textit{professional legislature binary}) + \beta_{11}(\textit{term limits}) + \beta_{12}(\textit{conflict scale}) + \beta_{13}(\textit{Trustee/Delegate scale}) + \beta_{14}(\textit{polarization binary}) + \mathcal{E}.$$

The results of these full ordinal regression models are shown in table 2.4. Note that these full models differ from the models in the previous section which used a composite dependent variable which was the median of all three forms of mature communication technology. The models in table 2.4 are full models with each individual mature communication technology as the dependent variable rather than a composite dependent variable.

Face-to-Face Communications

When all DPI control variables are utilized when analyzing the frequency of use of face-to-face communications, age, education, one chamber minority party status, and professional legislatures are all associated with increases in the frequency of use of face-to-face communications. Being black and in the other race category were both associated with decreases in the frequency of use of face-to-face communications. The largest effect found was the difference between the use of face-to-face communications by blacks and whites, with blacks being 54.2 percent less likely to choose "Use Hourly" over all other frequency of use categories for face-to-face communications than whites. The smallest effect found was the education measure where a

Table 2.4 Peer Frequency of Use Full Models, Odds Ratios

Frequency of Use	*Face-to-Face*	*Telephone*	*Letters*
Age	1.011*	0.996	1.004
Race (Compared to White)	Omitted Category: White	Omitted Category: White	Omitted Category: White
Black	0.458**	1.998*	1.173
Other	0.642*	0.945	1.044
Male	1.229	1.191	0.997
Education	1.044*	1.018	1.044*
Partyid (Conservativism)	1.013	1.048*	1.019
Minority Status (Compared to Majority Status)	Omitted Category: Majority Status	Omitted Category: Majority Status	Omitted Category: Majority Status
Minority One Chamber	1.483*	1.378	1.402
Minority Two Chambers	1.121	0.950	1.346
Minority Two Chambers and Governorship	1.050	0.968	1.049
Years in Office	1.005	1.027**	0.034***
Senate	1.273	1.358*	0.924
Gerrymandering (Composite)	1.003	1.002	1.999
Professional Legislature Binary	1.514**	1.589**	1.078
Term Limits	0.865	0.926	0.980
Conflict Scale	1.003	1.005	1.002
Trustee/Delegate Scale	0.999	1.001	1.003
Polarization Binary	0.905	1.036	1.050
Sample Size	1061	1060	1052
Pseudo R-Squared	0.019	0.018	0.012

* z ≤.05, ** z ≤.01, *** z ≤.001.
Self-Generated by author using Microsoft Word.

one-year increase in education was associated with a 4.4 percent increase in the likelihood that a legislator would choose "Use Hourly" over all other frequency of use categories for face-to-face communications. If a legislator's political party is in the minority in one chamber, they are more likely to use face-to-face communications than their majority counterparts. This result was not reproduced for telephone or hardcopy letter use. As expected, professional legislators communicate more with their peers via face-to-face meetings than do citizen legislators.

Telephone Communications

As shown in table 2.4, being black, increases in political conservativism, years in office, being a Senator, and being in a professional legislature are all associated with increases in the frequency of use of telephone communications. No variables were statistically significantly associated with decreases in the frequency of use of the telephone. The largest effect found was the difference between the use of telephone communications by blacks and whites, with blacks being 99.6 percent more likely to choose "Use Hourly" over all other frequency of use options for telephone communications than whites. The smallest effect size is years in office. A one-year increase in the time a legislator has been in office is associated with a 2.7 percent increase in the likelihood that a legislator will choose "Use Hourly" for telephone communications over all other frequency of use options, on average, all else equal. As expected, professional legislators communicate more with their peers via the telephone than do citizen legislators.

Hardcopy Letters

Legislator use of hardcopy letters is associated with three small, but statistically significant variables: years in office, gerrymandering, and being a Delegate. Each of these three variables is associated with small increases in the use of hardcopy letters. The largest effect size is years in office where a one-year increase in the time a legislator has been in office is associated with a 2.1 percent increase in the likelihood a legislator will choose "Use Hourly" for hardcopy letters over all other frequency of use options, on average, all else equal. The smallest effect size is associated with the composite gerrymandering score where a one-unit decrease in gerrymandering score (indicating an increase in gerrymandering) is associated with a 0.3 percent increase in the likelihood a legislator will choose "Use Hourly" for hardcopy letters over all other frequency of use options, on average, all else equal.

Mature Communication Technology Importance, Peer Communications

Bivariate ordinal logistic regressions were run for the importance of mature communications when communicating with peers, and DPI variables. The results of these frequency of use, ordinal bivariate regressions are summarized in the sections below. No literature has been located which discusses the importance of any particular communication technology to a state legislator, so the results in this section cannot be compared to existing research.

Face-to-Face Communications

With respect to face-to-face communications, legislator polarization and professional legislature status were both associated with increases in the importance of face-to-face communications, while the "other" race category and increases in a legislator's perception of constituent policy conflicts in their district were associated with decreases in the importance of face-to-face communications. Legislator race had the largest effect size, with legislators from the "Other" race category (non-whites and non-blacks) being 45.2 percent less likely to choose "Very Important" for face-to-face communications over all other importance categories, compared to whites, on average, all else equal. The smallest effect size is conflict scale, where a 1 percent increase in a legislator's perception of conflict in their district reducing the likelihood that they will select "Very Important" for face-to-face communications over all other importance options, on average, all else equal.

One of the more interesting results associated with the importance of face-to-face communications in these bivariate regressions is its link to polarization: The more importance a legislator places on face-to-face communications with their peers, the more likely they are to self-identify as strongly conservative or strongly progressive.

Telephone Communications

The variables associated with increases in the importance of telephone communications are years in office, being a Senator, both polarization measures, and both professional legislature measures. Both education and a legislator's perception of constituent policy conflicts in their district were associated with decreases in the importance of telephone communications. The largest effect size is the binary polarization measure where legislators who identified themselves as "Strongly Progressive" or "Strongly Conservative" were 64.4 percent more likely to identify telephone communications with their peers as "Very Important" over all other importance categories, compared to their more moderate peers, on average, all else equal. The smallest effect size is associated with the conflict scale measure whereby a 1 percent increase in a legislator's perception of conflict within their district is associated with a 0.7 percent decrease in the likelihood that a legislator will identify telephone communications with their peers as "Very Important" over all other importance categories, on average, all else equal.

Among the more interesting findings: Increases in legislator education are associated with decreases in the importance of using the telephone to communicate with other legislators. As with the importance of face-to-face communications, the more polarized a legislator, the more importance they place on using the telephone to communicate with their peers.

Hardcopy Letters

The statistically significant variables associated with increases in the importance of hardcopy letters include age, years in office, being black, and being more of a Delegate. Two gerrymandering scales, Schwartzberg and Convex

Hull, were associated with decreases in the importance of hardcopy letters. The largest effect size was once again race, where black legislators were 68.4 percent more likely to indicate that hardcopy letters were "Very Important" over all other importance categories, compared to their white peers, on average, all else equal. The smallest effect size was the Delegate/Trustee scale where a 1 percent increase in a legislator's evaluation of themselves as a Delegate is associated with a 0.3 percent increase in the likelihood that a legislator will find hardcopy letters "Very Important" compared to all other importance categories, on average, all else equal.

Peer Communication Technology Importance Models

Three models were created to determine the relationships between the importance of mature communication technology when used with peers, and DPI variables. As with the other importance models, no research is referenced with respect to the findings represented in these models since no research could be located which discusses the importance of communication technology when legislators communicate with their peers.

The use of basic, intermediate, and advanced models is a common social science standard, and has nothing to do with the fact that three mature communication technologies were analyzed. The basic model contains legislator demographic information, the intermediate model adds primary political and institutional variables, and the advanced model adds secondary institutional variables. Additionally, the results of these three models are not presented in a table, but rather, are discussed in the paragraphs following the models presented below:

Basic Model: $pimportmature = \beta_0 + \beta_1(age) + \beta_2(race) + \beta_3(male) + \beta_4(education) + \varepsilon$.

Intermediate Model: $pimportmature = \beta_0 + \beta_1(age) + \beta_2(race) + \beta_3(male) + \beta_4(education) + \beta_5(conservativism) + \beta_6(minority\ status) + \beta_7(years\ in\ office) + \beta_8(senate) + \varepsilon$.

Full Model: $pimportmature = \beta_0 + \beta_1(age) + \beta_2(race) + \beta_3(male) + \beta_4(education) + \beta_5(conservativism) + \beta_6(minority\ status) + \beta_7(years\ in\ office) + \beta_8(senate) + \beta_9(gerrymandering) + \beta_{10}(professional\ legislature\ binary) + \beta_{11}(term\ limits) + \beta_{12}(conflict\ scale) + \beta_{13}(Trustee/Delegate\ scale) + \beta_{14}(polarization\ binary) + \varepsilon$.

The three models all use a composite dependent variable for mature communications that is the median of each of the three individual mature peer communication technology importance variables.

The basic model is statistically significant for race (odds ratio = 1.727, $z \leq 0.01$). Interpreting race in the basic model for clarity, being black is associated with a 72.7 percent increase in the likelihood a legislator will select "Very Important" for the use of mature communications with peers over all other communication technology importance categories, ceteris paribus.

The intermediate model is statistically significant for race (odds ratio = 1.771, $z \leq 0.01$). Interpreting race in the intermediate model for clarity, being black is associated with a 72.7 percent increase in the likelihood a legislator will select "Very Important" for the use of mature communications with peers over all other communication technology importance categories, ceteris paribus.

The full model is statistically significant for conflict scale (odds ratio = 0.990, $z \leq 0.01$), Delegate scale (odds ratio = 1.004, $z \leq 0.05$), and being polarized (odds ratio = 1.481, $z \leq 0.01$). Interpreting a single variable in the full model for clarity, a legislator self-identified as being "Strongly Progressive" or "Strongly Conservative" is associated with a 48.1 percent increase in the likelihood a legislator will select "Very Important" for the use of mature communications with peers over all other communication technology importance categories, ceteris paribus.

Full Regression Model, Importance of Mature Communication Technology, Peers

In the previous section, full regression models were used to analyze the importance of mature communication technologies when legislators communicate with their peers. In this section, the importance of each mature communication technology is analyzed using full regression models of the form:

$$\text{Peer Mature Technology Importance} = \beta_0 + \beta_1(\text{age}) + \beta_2(\text{race}) + \beta_3(\text{male}) + \beta_4(\text{education}) + \beta_5(\text{conservativism}) + \beta_6(\text{minority status}) + \beta_7(\text{years in office}) + \beta_8(\text{senate}) + \beta_9(\text{gerrymandering}) + \beta_{10}(\text{professional legislature binary}) + \beta_{11}(\text{term limits}) + \beta_{12}(\text{conflict scale}) + \beta_{13}(\text{Trustee/Delegate scale}) + \beta_{14}(\text{polarization binary}) + \mathcal{E}.$$

The results of these full ordinal regression models are shown in table 2.5. As noted in previous importance regression models, no existing research is referenced on the importance of communication technology when used by legislators to communicate with other legislators since none could be found.

Face-to-Face Communications

When all DPI control variables are utilized when analyzing the frequency of use of face-to-face communications, only professional legislatures and

Table 2.5 Peer Importance Full Models, Odds Ratios

Race (Compared to White)	*Omitted Category: White*	*Omitted Category: White*	*Omitted Category: White*
Black	0.806	1.451	1.008
Other	0.799	1.278	1.395
Male	1.123	1.090	1.342
Education	1.045	0.988	1.154
Partyid (Conservativism)	0.986	1.346	0.987
Minority Status (Compared to Majority Status)	Omitted Category: Majority Status	Omitted Category: Majority Status	Omitted Category: Majority Status
Minority One Chamber	1.105	0.894	1.108
Minority Two Chambers	1.115	1.091	1.253
Minority Two Chambers and Governorship	0.986	0.941	0.962
Years in Office	0.975*	1.013	1.021**
Senate	1.245	1.513**	1.020
Gerrymandering (Composite)	1.000	0.999	0.997*
Professional Legislature Binary	1.469*	1.282	1.126
Term Limits	1.076	0.980	0.992
Conflict Scale	0.987**	0.992*	0.994
Trustee/Delegate Scale	1.001	1.003	1.004*
Polarization Binary	1.459*	1.222 *	1.122
Sample Size	1056	1048	1053
Pseudo R-Squared	0.022	0.010	0.020

* z ≤.05, ** z ≤.01, *** z ≤.001.
Self-Generated by author using Microsoft Word.

polarization are associated with increases in the importance of face-to-face communications. An increase in legislator detection of constituent policy conflict in their district is associated with a decrease in the importance of face-to-face communications. The largest effect found was the professional legislature measure, with professional legislators being 46.9 percent more likely to choose "Very Important" for face-to-face communications over all other importance categories, on average, all else equal. The smallest effect found was the conflict measure, where a 1 percent increase in legislator detection of constituent policy conflict in their district is associated with a 1.3 percent decrease in the likelihood that a legislator will choose "Very Important" over all other face-to-face importance categories, ceteris paribus.

Telephone Communications

As shown in table 2.5, being a Senator and being ideologically polarized are both associated with increase in the importance of telephone communications, while an increase in legislator detection of constituent policy conflicts in their district is associated with a decrease in the importance of telephone communications. The largest effect size was a legislator being a Senator, with Senators being 51.3 percent more likely to choose "Very Important" over all other importance categories, on average, all else equal. The smallest effect found was the conflict measure, where a 1 percent increase in legislator detection of constituent policy conflict in their district is associated with a 0.8 percent decrease in the likelihood that a legislator will choose "Very Important" over all other telephone importance categories, on average, all else equal.

Hardcopy Letters

The importance legislators assign to hardcopy letters are associated with three small, but statistically significant variables: years in office, gerrymandering, and being a Delegate. Each of these three variables is associated with small increases in the importance of hardcopy letters. The largest effect size is years in office where a one-year increase in the time a legislator has been in office is associated with a 2.1 percent increase in the likelihood a legislator will choose "Very Important" for hardcopy letters over all other importance options, on average, all else equal. The smallest effect size is associated with the composite gerrymandering score where a one unit decrease in gerrymandering score (indicating an increase in gerrymandering) is associated with a 0.3 percent increase in the likelihood a legislator will choose "Very Important" for hardcopy letters over all other importance options, on average, all else equal.

CONSTITUENT COMMUNICATIONS

Legislators were asked to rate the frequency of use of and the importance they assigned to mature communications with their constituents. Importance was defined for legislators as "Importance is related to the likelihood that you will respond favorably to a request received from a constituent via one of the communication technologies shown below" (followed by a list of all communication technologies examined in the survey). Frequency of use and importance categories and coding are identical to peer communications discussed in the previous section. table 2.6 highlights the frequency of use and importance of mature communication technologies when communicating with constituents.

Referencing table 2.6, when communicating with their constituents, legislators most frequently use written letters, followed by telephone calls and then face-to-face meetings. Legislators find face-to-face meetings most important when communicating with their constituents, followed by telephone calls and

Table 2.6 Frequency of Use and Importance Constituent Descriptive Statistics

Frequency of Use	*Face-to-face*	*Telephone*	*Written Letters*
Do Not Use	7	10	26
Use Annually	51	32	20
Use Monthly	333	230	147
Use Weekly	743	656	360
Use Daily	437	603	807
Use Hourly	43	116	251
Mean/Mode	3.04/3	3.28/3	3.65/4
σ	0.865	0.884	0.995
Importance			
Do Not Use	4	8	26
Not Important	4	5	20
Slightly Important	33	49	147
Moderately Important	71	122	360
Important	401	570	807
Very Important	1576	813	251
Mean/Mode	4.57/5	4.35/5	3.65/4
σ	0.737	0.843	0.995

Self-Generated by author using Microsoft Word.

then written letters. Legislators find face-to-face meetings most important, but they use written letters more frequently.

Inferential statistical analyses examining legislator communication technology frequency of use and importance when communicating with constituents have the same variables and coding as the analyses in the previous section covering peer communications. The only difference between this section and the peer section is that all of the questions in this section asked legislators about *constituent* communications instead of *peer* communications.

Mature Communication Technology Frequency of Use with Constituents

Bivariate ordinal logistic regressions were run for legislator frequency of use of mature communications when communicating with their constituents. The results of the frequency of use bivariate regressions are discussed in the sections below.

Face-to-Face Communications

The frequency of use of face-to-face communications with constituents is associated with years in office, senate, political party, gender, the Reock gerrymandering scale, one and two chamber minority status, and the Trustee/Delegate scale. Years in office, Senators, conservativism, being male, the

Reock gerrymandering scale, one chamber minority party status, and the Delegate scale are all associated with increases in the frequency of use of face-to-face communications with constituents. Only two chamber minority party status is associated with a decrease in the frequency of use of face-to-face communications. The largest effect size is related to Senate membership, where Senators are 71.4 percent more likely to choose "Use Hourly" over all other frequency of use categories for face-to-face communications with constituents than their Representative counterparts, on average, all else equal. The smallest effect size occurs in the relationship between the frequency of face-to-face communications with constituents and the legislator's perception of their Trustee/Delegate role. A 1.0 percent increase in a legislator's perception of their role as a Delegate is associated with a 0.4 percent increase in the likelihood that the legislator will choose "Use Hourly" for the frequency of constituent face-to-face meetings over all other frequency of use categories, on average, all else equal.

Among the more interesting results is the finding that Senators communicate more via face-to-face meetings than do Representatives. This result mirrors the findings for frequency of use of mature communication technology when communicating with peers. The number of constituents Senators represent may play a role in this finding. In all bicameral state legislatures, Senators represent more constituents than do Representatives. This relationship is true for all mature communication technologies.

Telephone Communications

Several DPI variables had statistically significant relationships with the frequency of use of telephone communications with constituents. These relationships include age, years in office, chamber, political party, gender, education, the Reock gerrymandering measure, one and two chamber majority/minority status, and the Trustee/Delegate scale. Age, years in office, Senate membership, conservativism, being male, the Reock gerrymandering measure, and the Delegate scale are all associated with increases in the frequency of use of telephone communications. Education and two chamber and governorship minority party status are all associated with decreases in the frequency of use of telephone communications with constituents. The largest effect size occurs in the relationship between telephone frequency of use and chamber. Compared to Representatives, Senators are 63 percent more likely to choose "Use Hourly" over all other frequency of use categories for telephone communications with constituents, on average, all else equal. The smallest effect size is the relationship between telephone frequency of use with constituents and the Trustee/Delegate scale. A 1 percent increase in a legislator's perception of their role as a Delegate is associated with a 0.6 percent increase in the likelihood that a legislator will choose "Use Hourly" for telephone conversations with constituents over all other frequency of use categories, on average, all else equal.

Among the more interesting results associated with telephone use is the finding that the more conservative the legislator, the more likely they are to use the telephone to communicate with their peers. This relationship is true for all mature communication technologies. In every case, conservative legislators are using mature communication technologies more than their more progressive counterparts.

Hardcopy Letters

Hardcopy letter frequency of use with constituents has statistically significant relationships with: years in office, chamber, education, race, gerrymandering, two minority chamber measures, professional legislature status, and the Trustee/Delegate scale. Years in office, Senate membership, conservativism, professional legislator status, and the Delegate scale are all associated with increases in the frequency of use of hardcopy letters. Education, the "other" race category, and two chamber and governorship minority status are all associated with decreases in the frequency of use of hardcopy letters. The largest effect size occurs between the frequency of hardcopy letter use with constituents and chamber. Senators are 61.8 percent more likely to choose "Use Hourly" over all other frequency of use categories for hardcopy letter frequency of use with their constituents than their Representative counterparts, on average, all else equal. The smallest effect size occurs with the Trustee/Delegate scale. A 1.0 percent increase in a legislator's perception of their role as a Delegate is associated with a 0.5 percent increase in the likelihood that a legislator will choose "Use Hourly" for hardcopy letter use with constituents over all other frequency of use categories, on average, all else equal.

The frequency of telephone use to communicate with constituents was associated with legislator ideological polarization. The more ideologically polarized the legislator, the more frequently they communicate with their constituents via the telephone. This relationship held and was statistically significant across all constituent mature communication technology frequency of use categories. Looking back at mature communication technology *peer* communications, this relationship held true for all frequency of use categories except hardcopy letters.

Constituent Communication Technology Frequency of Use Models

Three ordinal logistic regression models were created to determine the relationships between constituent frequency of use of mature communication technology and DPI variables. The use of basic, intermediate, and advanced models is a common social science standard, and has nothing to do with the fact that three mature communication technologies were analyzed. The basic model contains legislator demographic information, the intermediate model adds primary political and institutional variables, and the advanced model adds secondary institutional variables. Additionally, the results of these three models are not presented in a table, but rather, are discussed in the paragraphs following the models presented below:

Basic Model: $cfreqmature = \beta_0 + \beta_1(age) + \beta_2(race) + \beta_3(male) + \beta_4(education) + \mathcal{E}$.

Intermediate Model: $cfreqmature = \beta_0 + \beta_1(age) + \beta_2(race) + \beta_3(male) + \beta_4(education) + \beta_5(conservativism) + \beta_6(minority\ status) + \beta_7(years\ in\ office) + \beta_8(senate) + \mathcal{E}$.

Full Model: $cfreqmature = \beta_0 + \beta_1(age) + \beta_2(race) + \beta_3(male) + \beta_4(education) + \beta_5(conservativism) + \beta_6(minority\ status) + \beta_7(years\ in\ office) + \beta_8(senate) + \beta_9(gerrymandering) + \beta_{10}(professional\ legislature\ binary) + \beta_{11}(term\ limits) + \beta_{12}(conflict\ scale) + \beta_{13}(Trustee/Delegate\ scale) + \beta_{14}(polarization\ binary) + \mathcal{E}$.

The three models all use a composite dependent variable for mature communications that is the median of each of the three individual mature constituent communication technology frequency of use variables.

The basic model is statistically significant for race ("Other" race category) (odds ratio = 0.734, $z \leq 0.05$), gender (being male) (odds ratio = 1.320, $z \leq 0.01$), and education (odds ratio = 0.964, $z \leq 0.05$). Interpreting gender in the basic model for clarity, being male is associated with a 32.0 percent increase in the likelihood a legislator will select "Use Hourly" for mature communications with constituents over all other frequency of use categories, ceteris paribus.

When all three mature communication technology frequency of use variables (face-to-face, telephone, and hardcopy letters) are grouped together, overall, frequency of use of mature communication technology decreases as education increases. This is consistent with research that suggests Internet-enabled communication technology use decreases as education decreases (Y. Chen and Persson, 2002; Schleife, 2006; Selwyn, Gorard, Furlong, and Madden, 2003) although some research finds that education is not associated with the frequency of use of Internet-enabled communication technology. This said, and if education does play a role in the frequency of use of Internet-enabled communication technology, and if rates of communication are held constant, then a decrease in the use of Internet-enabled communication technology *should* precipitate an increase in the use of mature communication technologies.

The intermediate model is statistically significant for race ("Other" race category) (odds ratio = 0.704, $z \leq 0.5$), education (odds ratio = 0.956, $z \leq 0.01$), conservativism (odds ratio = 1.040, $z \leq 0.05$), one chamber minority party (odds ratio = 1.630, $z \leq 0.01$), years in office (odds ratio = 1.034, $z \leq 0.001$), and being a Senator (odds ratio = 1.770, $z \leq 0.001$). Interpreting the conservativism variable in the intermediate model for clarity, a one unit increase in a legislator's self-identified level of conservativism is associated with a 4.0 percent increase in the likelihood a legislator will select "Use Hourly" for mature communications with constituents over all other frequency of use categories, ceteris paribus.

The full model is statistically significant for education (odds ratio = 0.957, $z \leq 0.5$), conservativism (odds ratio = 1.041, $z \leq 0.05$), one chamber minority party (odds ratio = 1.551, $z \leq 0.01$), years in office (odds ratio = 1.033, $z \leq 0.001$), being a Senator (odds ratio = 1.995, $z \leq 0.001$), being a professional legislator (odds ratio = 1.354, $z \leq 0.05$), and acting more like a Delegate (odds ratio = 1.005, $z \leq 0.01$). Interpreting the Delegate variable in the full model for clarity, a 1 percent increase in a legislator's self-identified role as a Delegate is associated with a 0.5 percent increase in the likelihood a legislator will select "Use Hourly" for mature communications with constituents over all other frequency of use categories, ceteris paribus.

The relationship between the frequency of use of mature communication technologies and education is consistent across all three models. In all cases, as legislator education increases, the frequency of use of mature communication technologies decreases. This finding may be associated with the finding that more educated individuals tend to use computers less frequently than more highly educated individuals (Cutler, Hendricks, and Guyer, 2003; Schleife, 2006; Selwyn et al., 2003), leaving mature communication technologies as the only alternative for less educated legislators.

Full Regression Model, Mature Communication Technology Frequency of Use with Constituents

The previous sections analyzed and discussed mature communication technology relationships with constituents using bivariate regressions. In this section, the frequency of use of each mature communication technology is analyzed using a full regression model of the form:

Constituent Mature Technology Frequency of use $= \beta_0 + \beta_1(age) + \beta_2(race) + \beta_3(male) + \beta_4(education) + \beta_5(conservativism) + \beta_6(minority\ status) + \beta_7(years\ in\ office) + \beta_8(senate) + \beta_9(gerrymandering) + \beta_{10}(professional\ legislature\ binary) + \beta_{11}(term\ limits) + \beta_{12}(conflict\ scale) + \beta_{13}(Trustee/Delegate\ scale) + \beta_{14}(polarization\ binary) + \mathcal{E}$.

The results of this full ordinal regression model are found in table 2.7 for mature communication technology frequency of use.

Face-to-Face Communications

With respect to the frequency of use of face-to-face communications with constituents, minority party status in one chamber, years in office, and Senate membership are all associated with increases in the frequency of use of face-to-face communications with constituents. Minority status in two chambers is associated with decreases in the frequency of use of face-to-face communications. The largest effect found was the difference between the use of face-to-face communications by Senators and Representatives, with Senators being 85.6 percent less likely to choose "Use Hourly" over all other frequency of use categories for face-to-face communications than Representatives, on

Table 2.7 Constituent Frequency of Use Full Models, Odds Ratios

Frequency of Use	*Face-to-Face*	*Telephone*	*Letters*
Age	1.00	1.00	0.990
Race (Compared to White)	Omitted Category: White	Omitted Category: White	Omitted Category: White
Black	1.099	1.300	0.770
Other	0.882	0.899	0.633*
Male	1.282	1.363*	0.903
Education	0.987	0.952*	0.964
Partyid (Conservativism)	1.023	1.065**	1.013
Minority Status (Compared to Majority Status)	Omitted Category: Majority Status	Omitted Category: Majority Status	Omitted Category: Majority Status
Minority One Chamber	1.550*	1.318	1.418
Minority Two Chambers	0.658*	0.717	0.901
Minority Two Chambers and Governorship	1.086	1.043	0.811
Years in Office	1.021**	1.045***	1.031***
Senate	1.856***	1.809***	1.752***
Gerrymandering (Composite)	1.000	1.000	0.999
Professional Legislature Binary	1.274	1.288	1.331
Term Limits	0.945	1.040	0.866
Conflict Scale	0.999	0.996	0.995 (0.004)
Trustee/Delegate Scale	1.002	1.010***	1.004**
Polarization Binary	0.772	0.834	0.971
Sample Size	1058	1057	1057
Pseudo R-Squared	0.023	0.039	0.023

* $z \leq .05$, ** $z \leq .01$, *** $z \leq .001$.
Self-Generated by author using Microsoft Word.

average, all else equal. The smallest effect found was the years in office measure where a one-year increase in number of years a legislator is in office is associated with a 2.1 percent increase in the likelihood that a legislator would choose "Use Hourly" for face-to-face communications with constituents over all other frequency of use options, on average, all else equal.

Telephone Communications

As shown in table 2.7, gender, conservativism, years in office, Senate membership, and a 1 percent increase in acting like a Delegate are all associated with increases in the frequency of use of telephone communications. Only education is associated with a decrease in telephone communications. The largest effect found was the difference between Senators and Representatives, with Senators being 80.9 percent more likely to choose "Use Hourly" over all other frequency of use options for telephone communications than Representatives, on average, all else equal. The smallest effect size is related to a legislator seeing themselves as a Delegate. A 1.0 percent increase in a legislator's perception of themselves as a Delegate is associated with a with a 1.0 percent increase in the likelihood that a legislator will choose "Use Hourly" for telephone communications with constituents over all other frequency of use options, on average, all else equal.

One interesting finding that spans all three forms of mature communication technology used to communicate with constituents is that Senators use mature communication technology more than Representatives and the longer a legislator is in office, the more frequently they communicate with their constituents via a mature communication technology. These findings hold after controlling for all demographic variables available in the dataset.

Hardcopy Letters

Legislator use of hardcopy letters is associated with four statistically significant variables: the "other" race, years in office, senate membership, and the legislator's perception of themselves as a Delegate. Years in office, Senate membership, and an increase in a legislator's perception of themselves as a Delegate are all associated with increased use of letters to communicate with constituents. The "other" race category is associated with a decrease in the frequency of using hardcopy letters to communicate with constituents. The largest effect size is associated with Senate membership, where Senate membership is associated with a 75.2 percent increase in the likelihood a legislator will choose "Use Hourly" for hardcopy letters over all other frequency of use options, compared to Representatives, on average, all else equal. The smallest effect size is associated with legislators perceiving themselves as a

Delegate. A 1 percent increase in a legislator's perception of themselves as a Delegate is associated with a 0.4 percent increase in the likelihood a legislator will choose "Use Hourly" for hardcopy letters over all other frequency of use options, on average, all else equal.

Mature Communication Technology Importance, Constituent Communications

Bivariate ordinal logistic regressions were run for the importance legislators assign to communications when communicating with their constituents. The results of the importance bivariate regressions are discussed in the sections below. As noted in the peer communication category discussed previously, no reference to the concept of the importance of a mature communication technology was located, effectively eliminating the possibility of a discussion of existing research when presenting the findings below.

Face-to-Face Communications

The importance of face-to-face communications with constituents varies with age, both polarization measures, the "other" race category, two chamber minority status, both professional legislature measures, and with the conflict and Delegate/Trustee scales. Increases in the importance of face-to-face communications with constituents are associated with legislator polarization, two chamber minority party status, and the Delegate scale. Decreases in the importance of face-to-face communications with constituents are associated with age, the "other" race category, and the conflict scale. The largest effect size occurs between the importance of face-to-face communications with constituents and the two-chamber minority status variable. Legislators who are in the minority in both the upper and lower houses of their legislature are 60.2 percent more likely to choose "Very Important" for constituent face-to-face communications over all other importance categories, than their majority peers, on average, all else equal. The smallest effect size occurs between the importance of face-to-face communications with constituents and the Trustee/Delegate scale. A 1 percent increase in a legislator's perception of their role as a Delegate is associated with a 0.3 percent increase in the likelihood a legislator will choose "Very Important" for face-to-face constituent communications over all other importance categories, on average, all else equal.

Telephone Communications

The importance of telephone communications with constituents has statistically significant variations with respect to the polarization binary measure, the "other" race category, professional legislature status, and the Trustee/Delegate scale. Increases in the importance of telephone communications with constituents are associated with the polarization binary measure, professional legislature status, and the Delegate scale. Decreases in the importance of telephone communications with constituents are associated with the "other" race category. The largest effect size occurs with the polarization binary variable. Legislators who identify as "Strongly Progressive" or "Strongly Conservative" are 39.9 percent more likely to choose "Very Important" over all other importance categories for telephone communications with constituents, over their more moderate peers, on average, all else equal. The smallest effect size occurs between the importance of telephone communications with constituents and the Trustee/Delegate scale. A 1 percent increase in a legislator's perception of their role as a Delegate is associated with a 0.4 percent increase in the likelihood a legislator will choose "Very Important" over all other importance categories for telephone communications with constituents, on average, all else equal.

Hardcopy Letter Communications

Statistically significant relationships between the importance of hardcopy letters when communicating with constituents, and DPI variables include both polarization variables, gender, the "other" race category, both professional legislature measures, and the Trustee/Delegate scale. Increases in the importance of hardcopy letters when communicating with constituents are associated with both polarization measures and both professional legislature measures. Decreases in the importance of using hardcopy letters to communicate with constituents are associated with gender. The largest effect size occurs with the polarization binary variable. Legislators who identify as "Strongly Progressive" or "Strongly Conservative" are 57.5 percent more likely to choose "Very Important" over all other importance categories for hardcopy letter communications with constituents over their more moderate peers, on average, all else equal. The smallest effect size occurs between the importance of hardcopy letter communications with constituents and the Trustee/Delegate scale. A 1 percent increase in a legislator's perception of their role as a Delegate is associated with a 0.4 percent increase in the likelihood a legislator will choose "Very Important" for hardcopy letter communications with constituents over all other importance categories, on average, all else equal.

Constituent Communication Technology Importance Models

Three models were created to determine the relationships between the importance of mature communication technology when used with constituents, and DPI variables. The use of basic, intermediate, and advanced models is a common social science standard, and has nothing to do with the fact that three mature communication technologies were analyzed. The basic model contains legislator demographic information, the intermediate model adds primary political and institutional variables, and the advanced model adds secondary institutional variables. Additionally, the results of these three models are not presented in a table, but rather, are discussed in the paragraphs following the models presented below:

Basic Model: $cimportmature = \beta_0 + \beta_1(age) + \beta_2(race) + \beta_3(male) + \beta_4(education) + \mathcal{E}.$

Intermediate Model: $cimportmature = \beta_0 + \beta_1(age) + \beta_2(race) + \beta_3(male) + \beta_4(education) + \beta_5(conservativism) + \beta_6(minority\ status) + \beta_7(years\ in\ office) + \beta_8(senate) + \mathcal{E}.$

Full Model: $cimportmature = \beta_0 + \beta_1(age) + \beta_2(race) + \beta_3(male) + \beta_4(education) + \beta_5(conservativism) + \beta_6(minority\ status) + \beta_7(years\ in\ office) + \beta_8(senate) + \beta_9(gerrymandering) + \beta_{10}(professional\ legislature\ binary) + \beta_{11}(term\ limits) + \beta_{12}(conflict\ scale) + \beta_{13}(Trustee\ /\ Delegate\ scale) + \beta_{14}(polarization\ binary) + \mathcal{E}.$

The three models all use a composite dependent variable for mature communications that is the median of each of the three individual mature constituent communication technology importance variables.

The basic model is statistically significant for age (odds ratio = 0.989, $z \leq 0.05$) and the "Other" race category (odds ratio = 0.708, $z \leq 0.05$). Interpreting race in the basic model for clarity, being from the "Other" race category is associated with a 20.3 percent decrease in the likelihood a legislator will select "Very Important" over all other importance categories for the use of mature communications with constituents, ceteris paribus. Stating these results more simply, in the basic model, both age and a legislator's categorization as a member of the "other" race both are associated with decreases in the importance of a mature communication technology when used to communicate with constituents.

The intermediate model is statistically significant for age (odds ratio = 0.998, $z \leq 0.05$), gender (odds ratio = 0.749, $z \leq 0.01$), and being in a minority party in two legislative chambers (odds ratio = 1.496, $z \leq 0.05$). Interpreting age in the intermediate model for clarity, a one-year increase in legislator age is associated with a 0.2 percent increase in the likelihood a legislator will select "Very Important" over all other importance categories for the use of mature communications with constituents, ceteris paribus. Putting these results in common language, both gender (being male) and age are associated with decreases in the importance of a mature communication technology, while minority status in both legislative chambers is associated with an increase in the importance of mature communication technology.

The full model is statistically significant for gender (odds ratio = 0.770, $z \leq 0.05$), conflict scale (odds ratio = 0.990, $z \leq 0.01$), and the Delegate scale (odds ratio = 1.004, $z \leq 0.05$). Interpreting a single variable in the full model for clarity, a legislator detecting a 1 percent increase in the policy conflict among their constituents is associated with a 1.0 percent increase in the likelihood a legislator will select "Very Important" over all other importance categories for the use of mature communications with constituents, ceteris paribus. Put in common language, being male and the detection of more conflict in a legislator's district are both associated with decreases in the importance of a mature communication technology when used to communicate with a constituent and a legislator's perception of their role as a Delegate is associated with an increase in the importance of a mature communication technology when used to communicate with a constituent.

MATURE COMMUNICATION TECHNOLOGY IMPORTANCE, CONSTITUENT COMMUNICATIONS

The previous sections analyzed and discussed mature communication technology importance when communicating with constituents using regression models with an averaged dependent variable. In this section, the importance of each mature communication technology is analyzed using each mature communication technology as a dependent variable and a full regression model of the form:

Constituent Mature Technology Importance $= \beta_0 + \beta_1$ *(age)* $+ \beta_2$ *(race)* $+ \beta_3$ *(male)* $+ \beta_4$ *(education)* $+ \beta_5$ *(conservativism)* $+ \beta_6$ *(minority status)* $+ \beta_7$ *(years in office)* $+ \beta_8$ *(senate)* $+ \beta_9$ *(gerrymandering)* $+ \beta_{10}$ *(professional legislature binary)* $+ \beta_{11}$ *(term limits)* $+ \beta_{12}$ *(conflict scale)* $+ \beta_{13}$ *(Trustee/Delegate scale)* $+ \beta_{14}$ *(polarization binary)* $+ \mathcal{E}$.

The results of these full ordinal regression models are found in table 2.8.

Table 2.8 Constituent Importance Full Models, Odds Ratios

Importance	*Face-to-Face*	*Telephone*	*Letters*
Age	0.989	0.994	0.996
Race (Compared to White)	Omitted Category: White	Omitted Category: White	Omitted Category: White
Black	0.869	0.983	0. 524*
Other	0.871	0.717	0.700
Male	1.017	0.933	0.586***
Education	1.009	0.995	0.987
Partyid (Conservativism)	0.989	1.041	0.995
Minority Status (Compared to Majority Status)	Omitted Category: Majority Status	Omitted Category: Majority Status	Omitted Category: Majority Status
Minority One Chamber	1.141	1.083	1.129
Minority Two Chambers	1.397	1.341	1.261
Minority Two Chambers and Governorship	0.789	0.844	0.727*
Years in Office	0.988	0.997	0.994
Senate	1.087	1.087	1.356*
Gerrymandering (Composite)	1.000	0.999	0.998
Professional Legislature Binary	1.127	1.019	1.104
Term Limits	0.931	1.204	1.111
Conflict Scale	0.931**	0.994*	0.993
Trustee/Delegate Scale	1.002	1.003	1.003
Polarization Binary	1.207	1.004	1.223
Sample Size	1055	1048	1053
Pseudo R-Squared	0.015	0.010	0.020

* z ≤.05, ** z ≤.01, *** z ≤.001.
Self-Generated by author using Microsoft Word.

Face-to-Face Communications

With respect to the importance of face-to-face communications with constituents, only an increase in the amount of policy conflict a legislator detects in their district was statistically significant. A 1.0 percent increase in the constituent policy conflict a legislator detects in their district is associated with a 6.9 percent decrease in the likelihood that a legislator would choose "Very Important" for face-to-face communications with constituents over all other importance categories, on average, all else equal.

Telephone Communications

With respect to the importance of telephone communications with constituents, only an increase in the amount of policy conflict a legislator detects in their district was statistically significant. A 1.0 percent increase in the constituent policy conflict a legislator detects in their district is associated with a 0.6 percent decrease in the likelihood that a legislator would choose "Very Important" for face-to-face communications with constituents over all other importance categories, on average, all else equal.

Hardcopy Letters

Unlike the previous two mature communications where there was only a single statistically significant relationship, the importance legislators assign to hardcopy letters is associated with four statistically significant variables: being black, gender (being male), minority party status in two chambers and the governorship, and Senate membership. Only Senate membership is associated with increased importance of letters to communicate with constituents. The black race category, being male, and being in the minority party in both legislative chambers and the governorship are all associated with decreased importance assigned to the use of letters to communicate with constituents. The largest effect size is associated with being black. Being a black legislator is associated with a 47.6 percent decrease in the likelihood a legislator will choose "Very Important" for hardcopy letters over all other importance options, on average, all else equal. The smallest effect size is associated with legislators who are in the minority party in both legislative chambers and the governorship. Legislators who are in the minority party in both legislative chambers and the governorship are associated with a 27.3 percent decrease in the likelihood a legislator will choose "Very Important" for hardcopy letters over all other frequency of use options, on average, all else equal.

CONCLUSION

When communicating with their peers, legislators communicate most frequently via telephone conversations followed by face-to-face meetings, and finally by hardcopy letters, which are used on a monthly basis. With respect to the importance of mature communications with their peers, legislators find face-to-face meetings most important, followed by telephone calls and then hardcopy letters. When legislators communicate with their constituents, they use hardcopy letters most frequently, followed by telephone calls and, finally, face-to-face meetings most frequently. Legislators indicate that they prefer face-to-face meetings with their constituents, but they find that constituents do not engage them face-to-face. Legislators find face-to-face meetings most important for communicating with their constituents, followed by telephone use and, finally, written letters.

Interviews with legislators suggest that communication style is highly variable based on the personal preferences of legislators. This said, there are several themes which deserve to be summarized. First, legislators find face-to-face communications important for issues that they care about, or issues that their constituents care about. In a nod toward naturalness theory, legislators note that face-to-face communications allow them to read the body language of the person they are meeting with, which provides insights into how important the issue is, and, whether or not the legislator believes the person they are meeting with is telling the truth. Additionally, legislators note that face-to-face communications are both private and secure, allowing them to be free of FOIA requests. One of the most important aspects of face-to-face communications is that while legislators tend to prefer them, constituents do not, in general, request face-to-face meetings. Several legislators note that constituents seem to feel that legislators will be too busy to meet with them, or, that an individual constituent may feel that they are not important enough for the legislator to meet with them, so they do not even try to request a meeting. Legislators note that face-to-face meetings with constituents are a relatively rare occurrence.

With respect to the use of phones, legislators note a nearly universal distain to the use of automated phone calls that special interest groups use to connect constituents to legislators, noting that sometimes, constituents do not even realize they have been transferred from the phone call initiated by the special interest group to the legislator. Phone bank calls are as well received by legislators as boilerplate e-mails from constituents that all say the same thing. Legislators use phone calls most frequently because of their convenience and because constituents are typically free to talk on the phone. Additionally, legislators like hearing the emotion in the voice of their constituent, once again a nod toward naturalness theory. The ability to circumvent FOIA laws stands out as a reason why legislators like phone calls, although at least one legislator noted that phone logs are subject to FOIA requests.

Hardcopy letters are used for formal and important communications with constituents. Legislators feel that letters, such as a letter congratulating a new Eagle Scout in their district, are meaningful to their constituents.

The policy implications of the legislator communication behaviors highlighted in this chapter and framed by the control systems theory in chapter 1 fall into two of the four categories outlined in table 1.2: primary feedback loop characteristics. The four categories are: (1) feedback frequency, (2) crossover phase, (3) feedback duration, and (4) feedback path attenuation. As discussed in chapter 1, crossover phase and feedback durations were not measured as part of the data collection for this book, leaving feedback frequency and feedback path attenuation effects for discussion in the conclusion section of this, and all subsequent chapters.

With respect to feedback frequency of use when communicating with their peers, legislators indicate that the telephone is most frequently used, followed closely by face-to-face communications and then hardcopy letters. Frequency of use suggests that communicating with a legislator via the communication format they prefer will improve the likelihood that a legislator will make a favorable policy decision with respect to the individual they are communicating with. The importance of mature communication use with other legislators tells a different story; face-to-face communications are most important, followed by telephone use, and then hardcopy letters. Importance is a feedback attenuation effect; if one communicates with a legislator using a mature communication technology which is unimportant to that legislator, the chances of the legislator making a favorable policy decision with respect to the individual they are communicating with decrease.

Suggestive of the complexity surrounding legislator use of communication technology, the frequency of use of mature communication technologies when used with constituents is different than when used to communicate with peers. Legislators use written letters most frequently to communicate with their constituents, followed by the telephone, and finally, as noted in the interview section, least frequently used are face-to-face meetings. Once again, the importance of the communication technology tells another story. Legislators find face-to-face meetings most important, followed by the phone, with hardcopy letters being least important. In effect, legislators communicate with their constituents *most frequently* using hardcopy letters (the least important mature communication technology for legislators) and *least frequently* using face-to-face communications (the most important mature communication technology for legislators). This finding is important and speaks to a communication mismatch which decreases the significance of the communications that legislators receive from constituents. Simply put, this communication mismatch may very well be the most significant finding of this book. As later chapters will show, communication mismatches are associated with all communication technologies.

Multivariate regression models on peer and constituent use and importance of mature communication technologies produce many statistically significant results. Among them are the following: (1) senators use all forms of mature communication technologies to communicate with other legislators more than Representatives; (2) the longer a legislator is in office, the more frequently they use mature communication technologies to communicate with their constituents, even after controlling for age and education; (3) the more polarized a legislator is, the more important they find face-to-face and telephone communications; and (4) black legislators use face-to-face communications with their peers less than white legislators, but they use telephone communications with their peers almost twice as frequently as their white counterparts. There are many more significant findings in this chapter, each of which is associated with increases or decreases in the frequency of use of a communication technology, and with increases or decreases in the importance of a communication technology (path attenuation). The complexities associated with the frequency of use and importance of mature communication technologies sheds light on the importance of the critical frequency theory of policy system stability; communication technology matters in the policy process, and it matters much more than the r-squared values associated with the multivariate models presented in this chapter suggest.

NOTES

1. While both telephone calls and face-to-face meetings *can* utilize the Internet via voice over Internet protocol (VOIP) and online conferencing via Skype and WebEx, this chapter does not consider these technologies.

2. As one might imagine, qualitative interviews with legislators and their assistants in the other forty-nine states would have been prohibitively expensive and time consuming.

3. This network diagram is not presented due to format limitations; the network diagram is so large and complex that it is not readable within the constraints of the format of this book. A full copy of all network diagrams can be found on the Rowman and Littlefield website for this book.

Chapter 3

Mass-Media Communication Technologies

OVERVIEW

This chapter highlights the decline in both frequency of use and importance of some of the traditional forms of mass-media including radio, television, press releases, and town-hall style meetings. As state legislators turn toward Internet-enabled forms of communication technology to communicate with their constituents, they are turning away from traditional forms of mass-media. The consequences of this shift away from traditional mass-media are significant: long considered to be powerful from an agenda setting (and therefor policy) perspective, the traditional forms of mass-media may be diminishing in importance as the twenty-four hours per day, seven days per week Internet-enabled forms of communication become institutionalized and accepted forms of political communications. Both legislators and constituents are cutting ties with the more traditional forms of mass-media communications; and the result is a shift in power from a relatively small number of mass-media power players to a much more diverse number of "citizen reporters" who frequent online forums, dominate YouTube™ and the blogosphere, tweet policy in 140 characters, and dedicate countless hours to Facebook™ (which has been found to influence political behavior[1]) (Graber, 2009). The political landscape is shifting, and the political power that once belonged to CBS, NBC, and ABC in the years prior to the widespread use of the Internet now belongs to a new cadre of citizen news reporters; the hierarchy of power is being flattened. The consequences of this shift of power in the policymaking process are significant.

In 1956, C. Wright Mills (1956) drafted *The Power Elite*, a book which highlighted the individuals and entities in American society who held the reins of power. Listed among the powerful were "The Celebrities," individuals Mills describes as "The Names that need no further identification" (71–72). In 1956, the radio, television, and newspaper corporations in the United States had the power to make celebrities out of a chosen few (relatively speaking) and played a primary role in setting the policy agenda. Today, the Internet has the power to make celebrities out of virtually anyone, and to allow virtually anyone to set the policy agenda, even if it is only for a brief moment. As evidence of this power shift, one recent study found that Twitter™ use by the US Congress directed content in the *New York Times* newspaper (Shapiro and Hemphill, 2016).

For the purposes of this book, mass-media communications consist of press releases, town-hall meetings, television, and radio. Traditionally, mass-media communications are "push" communications from legislators to constituents (unidirectional) or they are communications from the public policy environment (such as television news outlets and radio talk shows) to legislators and constituents. This chapter focuses on the former category, examining how legislators use and value certain mass-media communications to communicate with their constituents. Importantly, as defined above, mass media is a public form of communication technology where mature communications covered in chapter 2 are considered a private form of communication technology.

Each of the following sections on constituent communications begins with descriptive statistics on the frequency of use and importance of mass-media communications and concludes with predictive linear and nonlinear models which examine the relationships between mass-media communications and DPI variables.

MASS-MEDIA COMMUNICATIONS

The influence of mass media on the policymaking process in the United States is well documented, and traditionally focuses on key aspects in the policymaking process such as agenda setting (Kingdon and Thurber, 2003; McCombs and Shaw, 1972; Schneider and Ingram, 1997), the effects of media on policy attitudes and behaviors (Graber, 2009; Kaid, 2004; Scheufele, 2002; Smith, 1989), and media impact on US elections (Graber, 2009; Pasek, Kenski, Romer, and Jamieson, 2006). The research in this chapter departs from the more traditional research into mass-media effects by focusing on how legislators use and value various forms of mass media to communicate with their constituents and peers.

Importantly, researchers are beginning to refer to Internet-enabled communication technology as mass-media communications,[2] but this chapter does not include any Internet-enabled communication technologies since these are covered in chapter 4. The analyses of legislator use of mass-media focus solely on legislator use of mass media to communicate with their constituents.

QUANTITATIVE ANALYSES

Demographic, Political, and Institutional Effects

Constituent Communications

Legislators were asked to rate the frequency of use and the importance they assigned to mass-media communications with their constituents. Frequency of use categories were "Do Not Use," "Use Annually," "Use Monthly," "Use Weekly," "Use Daily," and "Use Hourly." Importance categories were "Do not use," "Not Important," "Slightly Important," "Moderately Important," "Important," and "Very Important." Both frequency of use and importance categories were coded 0 through 5, where 0 was "Do Not Use" and 5 represented the highest frequency of use and importance. Table 3.1 highlights the frequency of use and importance of legislator use of mass-media communication technologies when communicating with their constituents.

Table 3.1 Frequency of Use and Importance of Mass Media, Descriptive Statistics

Frequency of Use	*Press Release*	*Town-Hall Meetings*	*Television*	*Radio*
Do Not Use	282	1065	805	1118
Use Annually	732	323	409	335
Use Monthly	468	154	242	106
Use Weekly	92	34	95	32
Use Daily	18	10	27	17
Use Hourly	8	3	6	4
Mean/Mode	1.29/1	0.496/0	0.831/0	0.434/0
σ	0.894	0.828	1.04	0.822
Importance				
Do Not Use	243	910	702	930
Not Important	73	143	140	270
Slightly Important	258	209	255	202
Moderately Important	298	134	203	94
Important	420	105	174	38
Very Important	257	41	68	23
Mean/Mode	2.87/3	1.03/0	1.49/1	0.781/ 0
σ	1.64	1.45	1.62	1.17

Self-Generated by author using Microsoft Word.

Referencing table 3.1, when communicating with their constituents, legislators most frequently use press releases followed by television, town-hall meetings, and then radio. With respect to mass-media importance, legislators find press releases most important when communicating with their constituents, followed by television, town-hall meetings, and radio. Unlike mature communication technologies, legislator use of mass-media aligns with the importance they assign to mass media; press releases are most frequently used and most important, television is second most frequently used and second most important, and so on.

One of the more interesting findings reflected in table 3.1 is that the traditional forms of mass-media communication technology examined are relatively unused; this result is reflected in the low mean and median numbers shown in table 3.1. Compared to mature and Internet-enabled communication technologies (discussed in the next chapter), mass-media communication technologies are, in general, the least used form of communication technology used to communicate with constituents. Some may find the lack of use of traditional mass media to be surprising, with radio, television, and town-hall style meetings relatively unused.

With respect to town-hall style meetings, researchers have noted that attendance in face-to-face town hall meetings has declined as "virtual" town-hall style meetings become more popular (Straus et al., 2013) with legislators by enhancing the ability to communicate with their constituents (Davis, 1999; Ines, 2015; Oleszek, 2007). Much like town-hall style meetings, radio and television use are also decreasing with the increasing popularity of the Internet (Dahlgren, 2005; Sala and Jones, 2012; Tufekci and Wilson, 2012), although some authors note that Internet-enabled communication technology is, to a certain extent, reliant on a symbiotic relationship with traditional mass-media communication technologies (Bimber and Davis, 2003).

Mass-Media Communication Technology Frequency of Use with Constituents

Bivariate ordinal logistic regressions were run for legislator frequency of use of mass-media communications when communicating with their constituents. The results of the frequency of use bivariate regressions are discussed below.

Press Releases

The frequency of use of press release communications with constituents is associated with age, senate membership, political party (conservativism measure), gender, being black, all five gerrymandering measures, conflict scale, and the Trustee/Delegate scale. Years in office, Senators, being black, all five gerrymandering scales, conflict scale, and the Delegate scale are all associated with increases in the frequency of use of press release communications with constituents. Increases in legislator age, increases in conservativism, and being male are all associated with decreases in the frequency of use of press release communications with constituents. The largest effect size is related to black legislators, where black legislators are 89.6 percent more likely to choose "Use Hourly" over all other frequency of use categories for press release communications with constituents than their white counterparts, on average, all else equal. The smallest effect size occurs in the relationship between the frequency of press release communications with constituents and the legislator's perception of the amount of policy conflict in their district. A 1 percent increase in a legislator's perception of policy conflict in their district is associated with a 0.1 percent increase in the likelihood that the legislator will choose "Use Hourly" for the frequency of constituent press releases over all other frequency of use categories, on average, all else equal.

In a rare occurrence in these analyses, all gerrymandering measures had highly ($p \leq .001$) statically significant relationships with legislator use of press releases; in all cases, lower levels of district gerrymandering were associated with increases in the use of press releases. If one assumes that decreases in gerrymandering are associated with more competitive districts, then an increase in press releases seems to make sense; however, it should be noted that this effect was less prevalent on the other forms of mass media. For example, decreases in gerrymandering were associated with increases in the use of radio in four of the five gerrymandering scales, decreases in the use of television in three of the five gerrymandering scales, and, with respect to town-hall meetings, decreases in gerrymandering were associated with an increase in the use of town-hall meetings in only one gerrymandering scale (the Reock scale).

Town-Hall Meetings

Several DPI variables had statistically significant relationships with the frequency of use of town-hall communications with constituents. These relationships include age, years in office, chamber, political party, gender, education, and the Reock gerrymandering measure. Years in office, Senate membership, being black, and the Reock gerrymandering measure are all associated with increases in the frequency of use of town-hall style communications with constituents. Age and conservativism are associated with decreases in the frequency of use of town-hall style communications with constituents. The largest effect size occurs in the relationship between town-hall frequency of use and black legislators. Compared to white legislators, black legislators are 143 percent more likely to choose "Use Hourly" for town-hall style communications with constituents over all other frequency of use categories, on average, all else equal. The smallest effect size is the relationship between town-hall meeting frequency of use with constituents and age. A one-year increase in a legislator's age is associated with a 1.2 percent increase in the likelihood that a legislator will choose "Use Hourly" for town-hall style communications with constituents over all other frequency of use categories, on average, all else equal.

Television

Television frequency of use with constituents has statistically significant relationships with years in office, chamber, education, conservativism, polarization, being black, the comprehensive, Polsby-popper, and Reock gerrymandering scales, term limits, and the Trustee/Delegate scale. All statistically significant measures are associated with increases in the frequency of use of television to communicate with constituents. The largest effect size occurs between the frequency of television contact with constituents and being a black legislator. Black legislators are 84.4 percent more likely to choose "Use Hourly" over all other frequency of use categories for television frequency of use with their constituents than their white counterparts, on average, all else equal. The smallest effect size occurs with the Trustee/Delegate scale. A 1 percent increase in a legislator's perception of their role as a Delegate is associated with a 0.3 percent increase in the likelihood that a legislator will choose "Use Hourly" for television frequency of use with constituents over all other frequency of use categories, on average, all else equal.

Radio

Radio frequency of use with constituents has statistically significant relationships with age, conservativism, education, race, gerrymandering, two minority chamber measures, professional legislature status, and the Trustee/Delegate scale. Years in office, Senate membership, conservativism, the polarization binary, being a black legislator, being in the "other" race category, all of the gerrymandering measures except Convex Hull, being in a professional legislature, being in a state with term limits, increases in a legislator's detection of policy conflict in their district, and the Delegate scale are all associated with increases in the frequency of use of radio communications with constituents. Age, conservativism, and being polarized are all associated with decreases in the frequency of use of radio communications with constituents. The largest effect size occurs with black legislators. Black legislators are 165 percent more likely to choose "Use Hourly" over all other frequency of use categories for radio frequency of use with their constituents than their white counterparts, on average, all else equal. The smallest effect size occurs with the Trustee/Delegate scale. A 1 percent increase in a legislator's perception of their role as a Delegate is associated with a 0.3 percent increase in the likelihood that a legislator will choose "Use Hourly" for radio communications use with constituents over all other frequency of use categories, on average, all else equal.

Mass-Media Communication Technology Importance Models

Three ordinal logistic regression models were created to determine the relationships between the importance of mass-media communication technology and DPI variables. The use of basic, intermediate, and advanced models is common a social science standard. The basic model contains legislator demographic information, the intermediate model adds primary political and institutional variables, and the advanced model adds secondary institutional variables. Additionally, the results of these three models are not presented in a table, but rather, are discussed in the paragraphs following the models presented below:

Basic Model: $cimportmedia = \beta_0 + \beta_1 (age) + \beta_2 (race) + \beta_3 (male) + \beta_4 (education) + \mathcal{E}.$

Intermediate Model: $cimportmedia = \beta_0 + \beta_1 (age) + \beta_2 (race) + \beta_3 (male) + \beta_4 (education) + \beta_5 (conservativism) + \beta_6 (minority\ status) + \beta_7 (years\ in\ office) + \beta_8 (senate) + \mathcal{E}.$

Full Model: $cimportmedia = \beta_0 + \beta_1 (age) + \beta_2 (race) + \beta_3 (male) + \beta_4 (education) + \beta_5 (conservativism) + \beta_6 (minority\ status) + \beta_7 (years\ in\ office) + \beta_8 (senate) + \beta_9 (gerrymandering) + \beta_{10} (professional\ legislature\ binary) + \beta_{11} (term\ limits) + \beta_{12} (conflict\ scale) + \beta_{13} (Trustee/Delegate\ scale) + \beta_{14} (polarization\ binary) + \mathcal{E}.$

The three models all use a composite dependent variable for mass-media communications that is the median of the sum of the four mass-media constituent communication technology importance variables (press releases, town-hall meetings, television, and radio).

The basic model is statistically significant for age (odds ratio = 0.974, $z \leq 0.001$), race (being black compared to white) (odds ratio = 3.298, $z \leq 0.001$), and gender (being male) (odds ratio = 0.712, $z \leq 0.01$). Interpreting race in the basic model for clarity, being black is associated with a 230 percent increase in the likelihood a legislator will select "Very Important" for mass-media communications with constituents over all other importance categories, ceteris paribus.

The intermediate model is statistically significant for age (odds ratio = 0.970, $z \leq 0.001$), race (being black compared to white) (odds ratio = 3.162, $z \leq 0.001$), gender (odds ratio = 0.671, $z \leq 0.001$), and being a Senator (odds ratio = 1.509, $z \leq 0.001$). Interpreting gender variable in the intermediate model for clarity, being a male legislator is associated with a 33 percent increase in the likelihood a legislator will select "Very Important" for mass-media communications with constituents over all other importance categories, ceteris paribus.

The full model is statistically significant for age (odds ratio = 0.969, $z \leq 0.001$), race (being black compared to white) (odds ratio = 2.931, $z \leq 0.001$), gender (being male) (odds ratio = 0.712, $z \leq 0.01$), gerrymandering (odds ratio = 1.007, $z \leq 0.001$), Senate membership (compared to Representatives) (odds ratio = 1.587, $z \leq 0.001$), and term limits (odds ratio = 0.680, $z \leq 0.01$). Interpreting the age variable in the full model for clarity, a one-year increase in legislator age is associated with a 3.1 percent decrease in the likelihood a legislator will select "Very Important" for mass-media communications with constituents over all other importance categories, ceteris paribus.

There are a number of findings which are consistent across all regression models: (1) Black legislators find mass media more important than white legislators, (2) Female legislators find mass media more important than male legislators, and (3) The older a legislator, the less importance they place on mass-media communication technology. Due to the dearth of research related to the importance of mass media as a function of race and gender, any attempt to locate these findings within the existing body of literature would be pure speculation.

The finding on the relationship between age and the importance of mass-media communication technology is somewhat troubling. One might expect that older legislators have a greater appreciation for mass media since it was the dominant form of mass communication prior to the widespread use of the Internet. The sign of this relationship did not change across any of the three models, suggesting that either the sign on this variable is correct, or, if the sign on this variable is incorrect, it is likely that a key explanatory variable is missing from the dataset. It is impossible to tell which of these two possibilities is correct.

Full Regression Model, Mass-media Communication Technology Frequency of Use

The previous sections analyzed and discussed mass-media communication technology use with constituents using bivariate regressions. In this section, the frequency of use of each mature communication technology is analyzed using a full regression model of the form:

Constituent Mass-media Technology Frequency of use = $\beta_0 + \beta_1(age) + \beta_2(race) + \beta_3(male) + \beta_4(education) + \beta_5(conservativism) + \beta_6(minority$

status) $+ \beta_7$*(years in office)* $+ \beta_8$*(senate)* $+ \beta_9$*(gerrymandering)* $+ \beta_{10}$*(professional legislature binary)* $+ \beta_{11}$*(term limits)* $+ \beta_{12}$*(conflict scale)* $+ \beta_{13}$ *(Trustee/Delegate scale)* $+ \beta_{14}$*(polarization binary)* $+ \mathcal{E}$.

The results of these full ordinal regression models are found in table 3.2 for mass-media communication technology frequency of use.

Press Release Communications

As shown in table 3.2, with respect to the frequency of use of press releases, race, Senate membership, and gerrymandering are all associated with increases in the frequency of use of press releases. Minority status in two chambers is associated with decreases in the frequency of use of press releases compared to the majority party. The largest effect found was the difference between the use of press releases by black legislators, with black legislators being 83.9 percent more likely to choose "Use Hourly" over all other frequency of use categories for press releases than white legislators, on average, all else equal. The smallest effect found was the gerrymandering measure where a one unit increase in gerrymandering score (less gerrymandering) is associated with a 1.2 percent increase in the likelihood that a legislator would choose "Use Hourly" for press release communications with constituents over all other frequency of use options, on average, all else equal.

Consistent with previous analyses, black legislators are using press releases more than their white counterparts and Senators are using press releases more frequently than Representatives; 2.5 percent of the variation in the use of press releases is explained by variations in the DPI independent variables.

Town-Hall Meeting Communications

Increases in the use of town-hall style meetings are associated with race, years in office, Senate membership, and gerrymandering. Age and term limits are associated with a decrease in the use of town-hall style communications. The largest effect found was the difference between black and white legislators, with black legislators being 87.1 percent more likely to choose "Use Hourly" over all other frequency of use options for town-hall communications compared to white legislators, on average, all else equal. The smallest effect size is related to state gerrymandering. A one unit increase in state gerrymandering (less gerrymandering) is associated with a 0.5 percent increase in the likelihood that a legislator will choose "Use Hourly" for town-hall style meetings with constituents over all other frequency of use options, on average, all else equal.

Once again, these findings are consistent with previous results; black legislators are using town hall meetings more frequently than white legislators,

Table 3.2 Constituent Frequency of Use Full Model, Odds Ratios

Frequency of Use	*Press Release*	*Town-Hall Meetings*	*Television*	*Radio*
Age	1.392	0.978***	0.985**	0.972***
Race (Compared to White)	Omitted Category: White	Omitted Category: White	Omitted Category: White	Omitted Category: White
Black	1.839*	1.871*	3.034	1.712
Other	1.292	1.054	0.919	1.211
Male	0.794	0.931	0.968	0.855
Education	0,992	1.045	1.004	0.993
Partyid (Conservativism)	0.984	0.994	1.079***	0.963
Minority Status (Compared to Majority Status)	Omitted Category: Majority Status	Omitted Category: Majority Status	Omitted Category: Majority Status	Omitted Category: Majority Status
Minority One Chamber	0.880	1.110	1.245	0.990
Minority Two Chambers	0.665*	1.181	1.022	0.625*
Minority Two Chambers and Governorship	1.109	1.135	1.252	1.020
Years in Office	0.989	1.037***	1.021**	1.004
Senate	1.335*	1.463**	1.803***	1.107
Gerrymandering (Composite)	1.012***	1.005*	1.008***	1.004*
Professional Legislature Binary	1.132	0.944	1.063	1.476*
Term Limits	0.972	0.605**	0.629***	1.202
Conflict Scale	1.005	1.009*	1.007	1.007
Trustee/Delegate Scale	1.005	0.999	1.002	1.004*
Polarization Binary	1.003	0.984	1.199	0.922
Sample Size	1048	1042	1039	1058
Pseudo R-Squared	0.025	0.030	0.033	0.032

* z ≤.05, ** z ≤.01, *** z ≤.001.
Self-Generated by Author using Microsoft Word.

and Senators are using town-hall meetings more frequently than Representatives. The composite gerrymandering variable suggests that the higher the gerrymandering score (i.e., the lower the amount of gerrymandering in a legislator's district), the more frequently a legislator will utilize town hall meetings. The effect size for the gerrymandering variable appears to be relatively small (0.5 percent change for a one unit change in gerrymandering score). Since the composite gerrymandering score has a 151.92 point range, this corresponds to an effect range of 75.96 percent change in the use of town-hall meetings as a function of gerrymandering.

Television Communications

Legislator use of television to communicate with constituents is associated with six statistically significant variables: age, partyid (a measure of conservativism), years in office, Senate membership, gerrymandering, and term limits. Partyid, Senate membership, and gerrymandering are all associated with increases in the use of television to communicate with constituents. Age and term limits are both associated with decreases in the use of television to communicate with constituents. The largest effect size is associated with Senate membership, where Senate membership is associated with an 80.3 percent increase in the likelihood a legislator will choose "Use Hourly" for television communications over all other frequency of use categories. The smallest effect size is associated with the number of years a legislator spends in office. A one-year increase in the number of years a legislator has been in office is associated with a 2.1 percent increase in the likelihood a legislator will choose "Use Hourly" for television communications over all other frequency of use options, on average, all else equal.

As with press releases and town-hall meetings, increased television use is associated with decreases in gerrymandering. This relationship seems intuitive since, all else being equal, an increase in the competitiveness of a district can reasonably be expected to increase a legislator's use of mass media.

Radio Communications

The final logistic regressions completed in this section are associated with radio communications. Legislator use of mass-media radio is associated with five statistically significant variables; these variables include age, minority status in both legislative chambers, gerrymandering, being a member in a professional legislature, and the legislator's perception of their role as a Trustee or Delegate. Gerrymandering, professional legislature membership, and their role as a Delegate/Trustee are all associated with an increase in the use of mass-media radio. Age and minority party status in two chambers are both associated with decreases in the use of mass-media radio. The largest effect size is associated with minority status in two chambers. Minority status in two chambers is associated with a 37.5 percent decrease in the likelihood a legislator will choose "Use Hourly" for mass-media communications over all other frequency of use options, compared to the majority party, on average, all else equal. The smallest effect size is associated with gerrymandering and Trustee/Delegate role. A 1 percent increase in a legislator's perception of themselves as a Delegate is associated with a 0.4 percent increase in the likelihood a legislator will choose "Use Hourly" for mass-media radio communications over all other frequency of use options, on average, all else equal.

As with all previous communication technologies, the frequency of use of radio is associated with gerrymandering, although radio use has the smallest effect size of any of the communication technologies analyzed.

Mass-Media Communication Technology Importance, Constituent Communications

The previous section analyzed the frequency of use of various mass-media communication technologies controlling for all DPI variables. In this section, the importance of each mass-media communication technology is analyzed using a full regression model of the form:

$$\textit{Mass-media Technology Importance} = \beta_0 + \beta_1(\textit{age}) + \beta_2(\textit{race}) + \beta_3(\textit{male}) + \beta_4(\textit{education}) + \beta_5(\textit{conservativism}) + \beta_6(\textit{minority status}) + \beta_7(\textit{years in office}) + \beta_8(\textit{senate}) + \beta_9(\textit{gerrymandering}) + \beta_{10}(\textit{professional legislature binary}) + \beta_{11}(\textit{term limits}) + \beta_{12}(\textit{conflict scale}) + \beta_{13}(\textit{Trustee/Delegate scale}) + \beta_{14}(\textit{polarization binary}) + \mathcal{E}.$$

The results of these full ordinal regression models are found in table 3.3.

Press Release Communications

As shown in table 3.3, with respect to the importance of press releases, age, gender, and the composite gerrymandering measure are statistically significant. Gerrymandering is associated with increases in the importance of press releases, while age and gender are associated with decreases in the importance of press releases. The largest effect found was the difference associated with gender. Males are 39.9 percent less likely to choose "Very Important" over all other importance categories for press releases than female legislators, on average, all else equal. The smallest effect found was the gerrymandering measure where one unit increases in gerrymandering score (less gerrymandering) is associated with a 0.7 percent increase in the likelihood that a legislator would choose "Very Important" for press release communications with constituents over all other frequency of use options, on average, all else equal.

Table 3.3 Constituent Importance Full Models, Odds Ratios

Importance	*Press Release*	*Town-Hall Meetings*	*Television*	*Radio*
Age	0.981***	0.973***	0.980***	0.972***
Race (Compared to White)	Omitted Category: White	Omitted Category: White	Omitted Category: White	Omitted Category: White
Black	1.157	2.617***	3.465***	2.000**
Other	1.230	1.438	1.177	1.199
Male	0.611***	0.939	0.960	0.867
Education	0.980	1.014	1.007	0.997
Partyid (Conservativism)	0.963	1.019	1.061**	0.962
Minority Status (Compared to Majority Status)	Omitted Category: Majority Status	Omitted Category: Majority Status	Omitted Category: Majority Status	Omitted Category: Majority Status
Minority One Chamber	1.072	1.062	1.232	1.157
Minority Two Chambers	0.783	1.491	1.315	0.786
Minority Two Chambers and Governorship	1.154	1.186	1.480**	0.968
Years in Office	0.988	1.029***	1.015	1.005
Senate	1.102	1.564**	1.906***	0.986
Gerrymandering (Composite)	1.007***	1.005*	1.006**	1.003
Professional Legislature Binary	1.062	0.860	0.873	1.263
Term Limits	0.819	0.559***	0.603**	1.090
Conflict Scale	0.997	1.004	1.001	1.008*
Trustee/Delegate Scale	1.002	1.001	1.003	1.003
Polarization Binary	1.142	0.985	1.231	1.008
Sample Size	1041	1036	1032	1048
Pseudo R-Squared	0.019	0.030	0.030	0.023

* z ≤.05, ** z ≤.01, *** z ≤.001.
Self-Generated by Author using Microsoft Word.

Town-Hall Meeting Communications

Increases in the use of town-hall style meetings are associated with age, race, years in office, Senate membership, gerrymandering, and term limits. Age and term limits are associated with a decrease in the importance of town-hall style communications, while race, years in office, and gerrymandering are associated with increases in the importance of town-hall style communications. The largest effect found was the difference between black and white legislators, with black legislators being 161.7 percent more likely to choose "Very Important" over all other importance options for town-hall communications when compared to white legislators, on average, all else equal. The smallest effect size is related to state gerrymandering. A one unit increase in state gerrymandering (less gerrymandering) is associated with a 0.5 percent increase in the likelihood that a legislator will choose "Very Important" for town-hall style meetings with constituents over all other importance categories, on average, all else equal.

The importance of press releases (and in fact, the importance of all mass-media communication technologies analyzed) increased with decreasing gerrymandering score. These results are consistent with the frequency of use of mass media as a function of gerrymandering score where mass media was used more frequently if a district was more competitive (i.e., less gerrymandering).

Television Communications

Legislator use of television to communicate with constituents is associated with seven statistically significant variables: age, race, partyid (conservativism), minority status in two chambers plus the governorship, Senate membership, gerrymandering, and term limits. Race, partyid, minority party in two chambers plus the governorship, and Senate membership are all associated with increases in the use of television to communicate with constituents. Age and term limits are both associated with decreases in the use of television to communicate with constituents. The largest effect size is associated with race. Black legislators are 246.5 percent more likely to choose "Very Important" for television communications over all other importance categories, on average, all else equal. The smallest effect size is associated with the gerrymandering scale. A one unit increase in the gerrymandering score for the state (less gerrymandering) is associated with a 0.6 percent increase in the likelihood a legislator will choose "Very Important" for television communications over all other frequency of use options, on average, all else equal.

Television was the only mass-media communication technology which had a statistically significant relationship with the conservativism measure (*partyid*). The more conservative a legislator, the more importance they placed on television.

Radio Communications

The final logistic regressions completed in this section are associated with radio communications. Legislator use of mass-media radio is associated with three statistically significant variables; these variables include age, race, and the conflict scale. Race and an increase in a legislator's perception of policy conflict in their district are both associated with an increase in the use of mass-media radio. Legislator age is associated with decreases in the use of mass-media radio. The largest effect size is associated with race. Black legislators are twice as likely to choose "Very Important" for mass-media communications over all other importance options, compared to white legislators, on average, all else equal. The smallest effect size is associated with the conflict scale. A 1 percent increase in a legislator's perception of the amount of policy conflict among their constituents is associated with a 0.8 percent increase in the likelihood a legislator will choose "Very Important" for mass-media radio communications over all other importance categories, on average, all else equal.

CONCLUSION

Mass media, once one of the most important forms of political communications for legislators, is, if the research presented in this chapter is any indication, no longer king of the hill, at least not for state legislators. Some researchers note that the Internet is a factor in the decreased importance of mass-media communication technologies (Kaid, 2004). Legislators use and value mass-media communication technologies less when compared with mature and Internet-enabled communication technologies. Legislators indicate that they use (mode = 0) and value/find important (mode = 1) mass-media communication technologies significantly less than all other forms of communication technology examined in this book. A full comparison of each communication technology is contained in chapter 5, although mass media is discussed less since peer communications were not examined for mass media as they were with mature and Internet-enabled communication technologies.

For mass-media communication technologies, legislators use press releases most frequently (mode = 1), followed by television (mode = 0), town-hall meetings (mode = 0), and radio (mode = 0). With respect to the relative importance of mass-media communication technologies, legislators find press releases most important (mode = 3), followed by television (mode = 1), town-hall meetings (mode = 0), and finally radio (mode = 0). The following sections summarize the statistically significant relationships between the frequency of use and importance of mass-media communication technologies and various DPI variables.

Full mass-media communication technology frequency of use regression models for each type of mass-media communication technology suggest that compared to white legislators, black legislators use press releases and town-hall meetings more frequently than white legislators, that one year of age reduces the use of town-hall meetings by 2 percent, television by 1 percent, and radio by 3 percent. For all mass-media communication technologies except radio, Senators use mass media more frequently than Representatives. The longer legislators are in office the more frequently they use town-hall meetings and television. The less gerrymandered a district, the more frequently legislators use all forms of mass-media communication technologies, and the more importance they place on them. Term limits are associated with a decrease in legislator use of town-hall meetings and television, and a corresponding decrease in the importance of these two mass-media formats. Finally, the more conflict legislators detect in their district, the more likely they are to use town-hall meetings (or vice versa) to communicate with their constituents.

Full mass-media communication technology importance regression models for each type of mass-media communication technology suggest that as legislators age, they find all forms of mass-media communication technology less important. Male legislators find press releases less important than do female legislators. Black legislators find all forms of mass-media communication technology except press releases more important than white legislators. Senators find town-hall meetings and television more important than Representatives, while greater years in office is correlated with an increase in the perceived importance of town-hall meetings. The more gerrymandered a legislator's district, the higher the importance assigned to all forms of mass-media communication technology. Term limits are associated with decreases in the importance of town-hall meetings and television.

With respect to the critical frequency theory, it is clear that mass media plays less of a role in the policymaking process (from a legislator perspective) than either mature communication technologies or Internet-enabled communication technologies. Additionally, the instrument used to gather the data in this book focused only on legislators using mass media to communicate with constituents. Because mass-media communications are expected to be both public and unidirectional (from legislators to constituents), their usefulness in the critical frequency theory is limited. Simply put, both mature and Internet-enabled communication technologies are bidirectional, and the constituent to legislator bi-directionality is an important part of the critical frequency theory.

This said, and if the 2016 US Presidential election is any example, there can be little doubt that *constituents* pay attention to mass-media outlets, both traditional (such as television) and non-traditional (such as Internet television). If legislators do not frequently use or value mass media for communicating with their constituents, then the messages from mass-media communications that constituents are receiving are coming from other sources. This said, it is reasonable to believe that the frequency of use and importance of mass media would be different if the instrument used to gather the data analyzed for this book focused specifically on campaigning. An important next step in developing the critical frequency theory more fully would be to determine the role that mass media plays in driving constituents to communicate with their legislators. This said, and thinking back to the communication technology mismatch using mature communication technologies, legislators communicate with their constituents *most frequently* using the least important mature communication technology and *least frequently* using the most important mature communication technology for legislators; it is possible that no matter how much mass media motivates citizens to communicate with legislators, the communications are ineffective.

NOTES

1. Well before the Cambridge Analytica/Facebook™ issue which led to the sharing of approximately 50 million Facebook™ profiles in an attempt to predict and influence choices at the ballot box during the 2016 US Presidential election.
2. Internet-enabled communication technology has mass-communication properties and is, in many ways, just as "mass media" as the traditional forms of mass media.

Chapter 4

Internet-Enabled Communication Technologies

OVERVIEW

Internet-enabled communication technologies (most frequently called ICT in the literature) consist of all communication technologies that use TCP/IP to transmit information between legislators and their peers and constituents. The Internet-enabled communication technologies examined in this chapter include e-mail, Twitter™, Facebook™, YouTube™, web pages, blogs, and text messaging.[1]

For the purposes of this book, Internet-enabled communication technologies can be defined as a combination of: (1) public communications in a public setting, (2) private communications in a public setting, and (3) private communications in a private setting. While mature communications are considered private communications and mass-media communications are considered public communications, Internet-enabled communication technologies are considered to be a mix of public and private communications.

One of the sections in this chapter examines the use of Internet-enabled communication technologies when used to communicate with other legislators. Although the instrument used did not specifically ask *why* legislators might use a public format to communicate with their peers, a review of the literature on legislator communications suggests that legislators use public communication formats to send messages not only to their constituents, but also to their peers, in a form of political posturing as a media strategy (Cook, 1989).

INTERNET-ENABLED COMMUNICATIONS

Internet-enabled communication technologies are the newest communication technologies examined in this book. US President Donald Trump's unusual use of Twitter as a bully pulpit has helped peak interest in how politicians use Internet-enabled communication technologies, but Twitter™ has long been a way to examine legislator behaviors, although the majority of the research located on Twitter™ focuses on a qualitative examination of the "tweets" rather than quantitative analyses of use behaviors. Current investigations into the use of Internet-enabled communication technologies by legislators includes research on Twitter™ (Cook, 2016; Evans, Cordova, and Sipole, 2014; Hwang, 2013; Shapiro and Hemphill, 2016; Straus et al., 2013), Facebook™ (Baltar and Brunet, 2012; Gulati and Williams, 2013; Khamis et al., 2012; Williams and Gulati, 2009), YouTube™ (Gulati and Williams, 2010; Khamis et al., 2012; Poell and Borra, 2012), and blogs (Bruns, Highfield, and Lind, 2012; Larsson and Moe, 2012; Wyld, 2007). At the time of the writing of this book, no existing research was located which examines the frequency of use and importance of various Internet-enabled communication technologies when legislators communicate with their peers and their constituents. This chapter will examine those relationships, beginning with peer communications.

According to Lathrop (2010), in the post-Internet age where access to information is rarely more than a mouse click or two away, the contact range of a constituent is greatly enlarged (p. 80) through Internet connections. No longer are constituents limited to influencing their own legislators, but they have the increased capability to influence other legislators through the expanded communication linkages offered by the Internet.

Acknowledgment of the importance of the impact of Internet-enabled communication technology (IECT) linkages can be found through a review of the literature on various forms of IECT linkages. Examples include the impact of online bloggers on the policy process (Hindman, 2010; Lathrop and Ruma, 2010; Lessig, 2006; Morozov, 2011; Oleszek, 2011; Shirky, 2008), the use and importance of Twitter™ to legislators (Cook, 2016; Evans et al., 2014; Graham, Broersma, Hazelhoff, and van't Haar, 2013; Hwang, 2013; Peterson, 2012; Poell and Borra, 2012; Straus et al., 2013), the use of Facebook™ in campaigns (Gulati and Williams, 2013) and congressional elections (Williams and Gulati, 2009), and the use of YouTube™ to effect political change (Helfert, 2017; Khamis et al., 2012; McNair, 2017).

In fact, there is so much information being sent to and from legislators because of the inexpensive nature of IECT (Bimber, 2003; Cook, 2016; Williams, Gulati, and DeLeo, 2013), that the concept of noise becomes important. The supply of information to legislators from citizens is virtually unbounded due to IECT technology, while the demand for information from constituents has decreased due to legislator use of IECT technology for their own research (Pole, 2005; West, 2014). Information supplied when it is not desired or requested is noise. As the level of noise increases, the ratio of signal (desired information) to noise (undesired information) decreases. Additionally, legislators lose control of political messages once they are released into the public sphere, a process McNair (2017) calls "Communication Chaos." Communication chaos adds to the noise that legislators must navigate. Nowhere is the concept of noise better seen than with constituent (and non-constituent) bombardment of legislators with boilerplate[2] e-mail; special interest groups plead with their members to send standardized e-mails on a particular policy topic to their legislators, yet legislators routinely ignore, and in some cases, actually become irritated at this influx of meaningless (to the legislator) communications.

An example of legislators turning to other informational sources because of excessive noise arises in the use of electronic mail (e-mail). The cost for constituents to lobby legislators via e-mail is extremely low since congressional mailing lists are freely available, allowing a single mouse click to send an e-mail to all legislators. The cost is so low in fact, that e-mail has become virtually meaningless as a form of communication to legislators (Shirky, 2008, 287; West, 2014), with many legislators using their staff to filter their e-mail by identifying important e-mail using an e-mail folder or by flagging important e-mails with an electronic "flag" to highlight what is important. This said, legislators find e-mail important for the "business" of legislating; legislators send e-mails to and receive from each other, their assistants, and legislative staff such as researchers, and so on.

According to Shirky (2008), Congress, in an attempt to stop non-constituents from influencing legislators, instituted a policy requiring the sender of an e-mail to include their full name and address. This requirement ended up failing because non-constituents simply used the Internet to look up addresses in a particular legislator's district and then copy and pasted them into their e-mails. As supporting evidence for Shirky's findings, in a survey of local elected officials conducted by Garson (2006), 25 percent reported receiving e-mail from constituents every day, while only 14 percent of those surveyed who reported receiving e-mail from constituents said they attached significant weight to e-mail. Research by West (2014) found that many legislators do not even read their own e-mail, allowing their Administrative Assistants and/or Chiefs of Staff to read, sort, prioritize, and triage important e-mails and then forward them to the legislator's personal e-mail account.

In *The Net Delusion* (Morozov, 2011), a book which examines some of the detrimental aspects of the use of the Internet as a linkage between citizens and governments, Morozov suggests that the abundance of information available on the Internet is not inherently supportive of the process of democracy and may in fact, disrupt relationships:

> The environment of information abundance is not by itself conducive to democratization, as it may disrupt a number of subtle but important relationships that help to nurture critical thinking. It's only now, as even democratic societies are navigating through this new environment of infinite content, that we realize that democracy is a much trickier, fragile, and demanding beast than we had previously assumed and that some of the conditions that enabled it [democracy] may have been highly specific to an epoch when information was scarce. (75)

Think back to the introduction to this book and the example of the time it took to distribute information in the 1700s to the time it takes to distribute information now; Morozov hints that the US form of representative democracy was optimized in a time when information was not readily available. One might infer from this paragraph that democracy may not, in fact, be optimized for a time when information is overly abundant.

It is fair to say that many, if not all, new communication technologies are ultimately adopted by legislators and constituents to communicate with each other, and the factors which determine communication technology adoption are themselves complex and include incumbency status, political party, legislative professionalism, among others (Williams et al., 2013). Each new communication technology adds another way of linking constituents with legislators (and legislators with each other) and, importantly, impacts existing linkages between constituents and legislators. For example, Kindra et al. (2014) note that traditional political party linkages to voters are decreasing in strength as new two-way communication technologies come into existence. In a second example, legislators indicate that constituents transitioned away from using phone calls and hardcopy letters as e-mail use became widespread (West, 2014).

New communication technologies not only build additional legislator–constituent linkages and impact the use and importance of existing linkages, but also influence the very nature of the communication itself. For example, researchers find that older legislators use Internet-enabled communication technologies less than do younger legislators (Greenberg, 2012; West and Corley, 2016), yet younger constituents use Internet-enabled communication technologies more frequently than older constituents (Liikanen, Stoneman, and Toivanen, 2004; Mergel, 2012; Palfrey and Gasser, 2013). This disconnection between older legislators and younger constituents could reduce the representation of a younger generation of constituents who are coupled to an older legislator. What are the impacts of this communication mismatch on the policymaking process? This question is explored in this book and conclusions drawn in the final chapter.

Policymaking linkages are more than simple connections between policymaking system outputs to the public policy environment and vice versa. As we will see in later chapters, communication technologies that comprise the various connections between policymaking systems and the public policy environment actually shape the policymaking process through changes in human behavior associated with the communication technology used. For example, some older and more experienced legislators eschew social media communications by their constituents, sometimes at the expense of their own reelection.

As demonstrated in the various chapters of this book, the frequency of use and importance of communication technologies vary as a function of DPI variables surrounding the policymaking process. These behavioral changes and the relationships with DPI variables have policy implications beyond those historically attributed to policy feedback paths and linkages. The examination of influences of communication technologies on the policymaking process using natural science concepts in the following section shed some light on these policy feedback paths and linkages. This approach is not particularly innovative, but rather, is an extension of previous work completed by policy process theorists. Using a natural science approach on social science topics has been eschewed by social science researchers such as Robert Dahl (1947) and embraced by others including Ludwig von Bertalanffy (1950), who wrote the seminal work *An Outline of General System Theory*, which argued that complex relationships can be modeled by breaking them into various interconnected systems. This book follows von Bertalanffy's general system theory approach by using the natural science concept of control systems engineering to model and evaluate the effects of communication technology on the policy process in the United States.

With an overview of Internet-enabled communication technology literature complete, focus now turns toward interviews with Arizona state legislators on the topic of Internet-enabled communication technologies.

LEGISLATOR INTERVIEWS

E-mail

E-mail was the single most grounded and connected communication technology discussed during interviews with legislators and legislator assistants. E-mail network analysis highlights clusters around six primary themes: bulk e-mail, e-mail efficiency, e-mail naturalness, e-mail risks, e-mail benefits, and behaviors associated with e-mail.

These themes appear as central nodes with second pass quasi In Vivo codes clustered around them. The quasi In Vivo second pass codes are supported by quotes from transcribed interviews. In general, there are many connections between these nodes that describe the relationships.

Although the individual network diagrams are too complex to be printed in detail (but are contained on Rowman and Littlefield's website for this book), they can be explained relatively easily. For example, connected to the network node "Behavior" for e-mail use is the code "Legislator Concerned More About Risk Than Benefit" with the association "is a part of." In other words, a legislator being concerned more about the risks of e-mail than the benefits of e-mail are a part of legislator behavior. In turn, the "Legislator Concerned More About Risk Than Benefit" node is connected to both the "e-mail Risks" and "e-mail Benefits" nodes via "is associated with" links, which in turn are connected to the coded risks and benefits that legislators associated with e-mail via "is a property of" links. The individual risks and benefits of e-mail are grounded to quotes in the interviews. Using e-mail risks as an example, the risk "No Control Over Information Once Released" is grounded by four quotes from interviews. E-mail behavior is grounded in 942 quotes and 95 individual codes. In the following sections, each of the six primary e-mail themes will be expanded and examined in detail. Where appropriate, links to other to other themes will be examined.

BULK E-MAIL

Bulk e-mail is e-mail that occurs in large numbers and has identical text other than minor personalization differences. In general, legislators indicate that they receive relatively large numbers of bulk e-mail which has little impact on them. A representative quote from legislator E:

> Yeah, I mean basically somebody . . . it's sent to, whatever the Sierra Club or . . . any NRA, they have organized groups, they type their name in and then it generates and e-mail to me, sent from them, but it's the same text . . . . And I consider those more to be public opinion polls, extremely non-random, extremely unreliable, so they really mean very little to me.

Two legislators suggested that constituents who use bulk e-mail often have little "clue" what they are signing.

Legislator L:

> I don't think that people that are actually putting their name on a—, on, a—, a form letter has any clue of what they're signing or not

Legislator K:

> They don't have a clue. They're just a member of a group and the leadership of the group has said we're going to oppose this and we need to get as many people to join us in the opposition as we can. And so I say I will just, from time to time, if I have a few minutes, I'll just pick one of those. I don't even, I have no clue. And I will send this. And it's amazing how often I get no response whatsoever from that person. Because that's the last thing they expected was some sort of actual, thoughtful response.

Assistant H on the topic of bulk e-mail:

> I would say, yes, it does. Like, for example, the form e-mail has very little impact at all on how the legislators is gonna vote.

Bulk e-mail is one of the most frequent forms of communication received by legislators from constituents, but, as can be seen from the interview data, is of little importance. None of the legislators had any comments with respect to bulk e-mail suggesting that it in any way changed their floor voting behavior. Bulk e-mail is identified as the most common constituent communication by legislator assistants and legislators, but it is also identified as the least effective communication technology in common use, an example of a communication mismatch. Mismatch between constituent communication technologies and the communication technologies deemed most important by legislators is one of the most important unanticipated results of this study and will be discussed further in the final chapter of this book.

E-MAIL EFFICIENCY

A second common thread in interviews was the concept of e-mail efficiency. Here is legislator E's discussion of efficiency:

> And then second, just based on sheer volume of course, is e-mail. Again, qualitatively it's pretty poor in terms of exchange of info, although I like the efficiency of e-mail; people don't expect you to do three minutes of BS how you're doing before you get to the point. e-mail's very efficient, to the point, cryptic is fine, and that's it, so I think it's . . . it's probably . . . to dramatically increase the flow and quality of efficient communication, so that's a blessing for it.

Codes associated with e-mail efficiency include convenience, quick response time, mass communications (quantity of information delivered at one time), and time savings. Here is legislator L on time savings:

> So I'm—, I'm more of the old school. Now I can tell you there's legislators down here that'd just as soon get it over the e-mail or get it on a text, or Facebook™ or whatever. Because I think they don't have to engage. You know, once it's there it's—, you look at it and you can send a quick message. And so it saves time.

E-mail efficiency is categorized as a benefit by legislators. As discussed in the quantitative results presented previously in this book, it is likely that the efficiency of e-mail is one of the major factors that raised its importance above that which naturalness theory would predict. Notice that legislator L indicates a reason that e-mail is efficient—legislators do not have to engage. Importantly, legislator L is indicating that *other* legislators prefer e-mail, but he is "old-school" and prefers face-to-face or telephone (as he indicates earlier in the interview).

E-MAIL NATURALNESS

Naturalness theory suggests that e-mail is not very "face-to-face" like. Interviews with legislators and staff produced codes related to naturalness theory suggesting that this is indeed the case.

Legislator L sums up his feelings on the impersonal nature of e-mail:

> e-mails, you can say whatever you want to in e-mails and it's just words on paper that really doesn't give personal effect, that I think. Same with a text message. You know, they're so short now because everybody is trying to be in a hurry to—they may give you some pertinent information real quick but it doesn't give you the detail that I like in-, in personal-, personal conversation.

Naturalness theory suggests that a phone call is more natural than an e-mail. Legislator L touches on this topic when he indicates a preference for a telephone call so that he can ask questions in a full-duplex (real-time interactive) mode:

> And gives me something that I can take to the people that I need to in order to start resolving it. So I would prefer—you can alert me with an e-mail but I would prefer to have a hard copy letter so that I have it in hand. Or a telephone call so that I know and can ask some questions about, "Well, how is this affected?" or "What do I need to do?" or how?" You know?

Legislators note that one of the risks of e-mail is that it is difficult to gage reactions and emotions in an e-mail. Here is legislator H on the topic:

> Well, face-to-face always gives me the benefit of knowing . . . reading their personalities or aura or whatever it is about them; how they react to my statements,

> much better than e-mail. On the phone you can do it a little bit, you know, but face-to-face always seems to work better as far as being able to read their reactions to an issue. Usually they give you both sides of an issue, and they'll tell you the pros and the cons, but some lobbyists don't, you know; they only say this is how it'll benefit you, and if you don't ask how on the negative side of it, they aren't going to tell you. So you have to probe and you know, you have to ask questions; you have to generate possible concerns that might come out of it, of a bill passing, so . . . . You can hear a bit, yeah, of what they're. . . . you know, by the tone of their voice and by their reaction to your questions, but in e-mail there's no personality there.

The above interviews indicate that e-mail does not offer the same amount of information that face-to-face communications offer, such as emotions and instant reactions to the topic being discussed. Legislator H notes that e-mail lacks personality. This is an important concept, and naturalness theory predicts that unnatural communications increase ambiguity (Kock, 2007) , and this ambiguity (and other factors associated with unnatural communications) leads to risks for legislators. Legislator H notes that some lobbyists offer only partial information, and that by being able to see them face-to-face, or hear the tone of their voice over the phone, the legislator can pick up on any deception.

E-MAIL RISKS

A common theme in the risks that legislators associate with e-mail is the speed at which it can be released. Listed as both a benefit and a risk, according to legislators, sending quickly without thinking can kill legislator careers. Legislator E:

> The other of course is, it's too easy to quickly send an e-mail, so you might not think. . . . You know, either you're driven by emotion or you didn't think out an issue properly; you made your response that you later probably shouldn't have made—that's another issue. And of course, there's always the possibility of reply all and totally screwing yourself.

Legislator G, having had his career almost derailed by information he made public before he was a legislator, notes the dangers of the quick reaction and combines it with the peril of the permanence of digital information:

> I think the risks are that to the extent that you're impulsive or you react quickly, or fly off the handle, you might say, to the extent that that's documented digitally; that can come back and haunt you in a campaign—those things get replayed and replayed and replayed in campaigns. So I find my personality is

> one of let's be thoughtful about what we say before we say it; let's be careful what we say. This isn't about me being the center of attention; this is about being a reasonable statesperson, thoughtful about what you say and what you do. Live your life in a conservative way, and so I think to some extent, if you get too carried away with the wonderful digital media tools we have—that we did not have ten, fifteen years ago—it's entirely possible that you might try to make yourself kind of a media icon—a digital media icon—which is up there chatting all the time, talking all the time and maybe saying too much

The theme of the risks associated with the permanence of digital information is echoed by other legislators:

Legislator D:

> You know anything you write, whether it's in an e-mail or physically, is there in perpetuity. And so, you know, if you don't have a firm grasp on what you think before you communicate it, it probably wouldn't be smart to do that. So, I try and think that through before I communicate in written communication.

Legislator H responding to a question about e-mail risks: "You have to be careful what you say in an e-mail anywhere, because they're public and they can request them."

Legislator H, raising the issue of Freedom of Information Act (FOIA) laws, points directly to a legislator behavior precipitated by FOIA and expressed by six of nine legislators. Legislator B expresses this behavior:

> Well there is an understanding within the legislative process that we organize ourselves within partisan frameworks but if it is party work or things that are exclusively unrelated to public policy but partisan in nature, can they work for another candidate, we never want to use the public communication tool for that so any communication in this setting has risk because it is subject to public scrutiny. Anything you put out that goes on to a Facebook™ or even an e-mail, can be released, that's not the right word, but forwarded on to others so you don't have control over where it goes so you have to make decisions, if you are in a private conversation with somebody you don't want to do it over legislative e-mail.

In effect, legislators turn to private communication technologies to bypass the dangers associated with FOIA disclosure. Legislator K on the subject of FOIA and then extending the FOIA risk to encompass yet another risk legislators associate with e-mail: a lack of distribution control:

> So, if you think at all that how you respond to an e-mail is somehow private, forget it. And I'm not talking about the fact that they can go in and make public records request and get your e-mail. I'm just talking about the fact that you can

> be assured and there is a few times that I was surprised by that. I thought I was just responding to John Doe, you know, in my district and then I discover that, you know, that's disseminated far and wide and sometimes will regret that. So, learned that like everybody else, after a few hard knocks.

Legislator H, tying communication risks to benefits by noting he is more concerned about risks than benefits, offers a segue to the benefits associated with e-mail:

> I'm concerned about risk, more than benefits, you know. Being in the legislature and being in public office and people taking pot shots at you, you know, depending on your position on issues, you take . . . you know, the risk is there.

Legislators understand the risks associated with e-mail, especially their official public e-mail, and work to find ways around the risk. As will be shown in future sections, all communication technologies examined were associated with risks; however, the more natural the communication technology, the fewer the risks legislators associated with it. Although legislators associate risks with e-mail, they also identify specific benefits (beyond efficiency). The next section examines these benefits.

E-MAIL BENEFITS

Although a majority of the benefits legislators associate with e-mail are related to its efficiency, many legislators focused on the mass-communication capabilities of e-mail as a benefit. Legislator L sums this up succinctly: "Well, I think-, I think benefits is that you can put-, you can blast out a message to a lot of people in one stroke key." Transparency is a benefit identified by Legislator G:

> The benefit is that the elected representatives are more transparent and what they're doing, what they're working on, what they're thinking, that's the benefit. The voters can see who they've elected. It used to be they'd come down here in smoke-filled rooms, back room deals, you didn't know, and then when there's an election cycle, you just know what they decide to tell you; now things are much more transparent, which I think is a good thing.

Assistant B suggests that people are more candid when they send e-mails, and she sees this as a benefit: "The benefits is, um, people tend to be more candid, maybe more so when they e-mail." Legislator K noted that e-mail allows legislators to communicate with each other during the third reading of a bill, when legislators are not allowed to leave their seats:

> When I was in the House, and on third read of bills in the House, you know, you can't leave your chair. Once you're in the third read you have to stay there. Now, you can e-mail somebody or you can text another member, you know, across the chamber, you know: How are you going to vote on this one? Something like that.

When legislators were asked about the risks and benefits of communication technology, e-mail was the single most risky communication technology based on interview coding network diagrams. Legislators seemed to be most concerned with the risks associated with the permanence of information associated with written, audio, and video technologies. Linked to these concerns is the inability to control information in these formats once it has been distributed. In effect, legislators are aware that anything which is recorded becomes part of the public domain, and information can rarely, if ever, be contained once it is released, whether that release was intentional or not.

E-MAIL-ASSOCIATED LEGISLATOR BEHAVIORS

Many of the behaviors, such as using private e-mail to circumvent FOIA laws and the concept that legislators, are more concerned about communication technology risk than communication technology benefits that have been covered in previous sections, and other behaviors are self-explanatory from looking at the network node quasi In Vivo code name.

Age-related e-mail behaviors align nicely with the expectations set in chapter 2—namely that older individuals are likely to use technology less. The following interview data support the age-related regression models discussed in the quantitative results earlier in this book. Here is Staffer F's perspective:

> —so we have a lot of older constituents. A lot of them don't have e-mail or don't like to use e-mail. These are older folks, mm-hmm. I think . . . . From what I've seen and—, and in dealing with—, with the older generations, they like letters. They like letters because that's—, that's what they're used to. And personally too, it's more personal.

Assistant F went on to tell a story about an eighty-three-year old female constituent who calls her once a week just to talk about the things going on in her life and current events. IT A contributes to the discussion regarding age related e-mail behavior in this passage:

> There's a wide range of behaviors out there; there is from older members that don't really want to use their technology at all—they tend to resist it still, they

> have their assistant do all the communication through e-mail, etc.—and then there is members in the middle that use the technology, but they don't demand and really embrace it, and then we have a group of younger members that are really demanding that they have greater communication tools that's available, and it does not seem to be each specific—I would say that the assistant would have to filter out what's necessary correspondence or not, and then either draft up a letter or e-mail to the constituent on behalf of the member. Most members are literate enough that they can do e-mails themselves, currently; they may not want to, that's the issue.

In this passage, IT A categorizes three types of legislators who use communication technology, all delineated by age, and then notes that older members (legislators) have their legislator assistant filter and relay e-mail information to the older legislator, while younger legislators "demand" the best communication technology that is available.

All three IT staff interviewed perceived legislator behaviors associated with communication technology as being related to age, exclusively. When specifically questioned about gender or education or other DPI variables they notice impact legislator behavior associated with communication technology, only age was mentioned.

Here is legislator B on the topic of older constituents:

> Younger folks communicate with us by e-mail. They, the older folks more experienced folks who didn't grow up in a digital world are on a different edge of that digital divide we talked about and so they tend to call. With the exception of some of the retired school teachers in the district who were early adopters in the digital technology. I hear from some of my retired teachers and teacher friends who were in their 80s and 90s.

Legislator B notes that she has seen some exceptions to the rule that older constituents prefer the phone.

Constituent Selective Behavior

One of the more interesting legislator behaviors related to e-mail is coded as "Constituent Selective Behavior." Constituent Selective Behavior (CSB) occurs when legislators engage in acts that either limit or enhance their exposure to certain types of constituents as a normal function of their day-to-day communications acting in their capacity as a legislator. The definition of CSB specifically excludes campaign activities as such activities that typically involve enhanced exposure to certain types of constituents (usually their supporters). CSB is associated with two behaviors: constituent selective exposure and constituent challenge avoidance. Constituent selective exposure can be defined as legislator actions that seek to enhance their exposure to certain types of constituents as a normal function of their day-to-day communications acting in their capacity as a legislator. Constituent challenge avoidance occurs when legislators seek to reduce their exposure to certain types of constituents as a normal function of their day-to-day communications acting in their capacity as a legislator.

CSB is one of three most important unanticipated results[3] that will be more thoroughly discussed in the final chapter of this book. With respect to CSB behaviors, e-mail was grounded in CSB behavior in thirteen coded exchanges in interviews with legislators. In the following example, legislator J responds to a question asking how she filters communications from constituents:

> It depends on the issue. If they don't include their address or if they include their address and it is outside of my district I just hit delete unless it is an issue I agree in, so for instance I agree in protecting second amendment rights so I am in agreement with gun rights and so if I have people that e-mail me from throughout the state saying protect my gun rights, I am going to save their e-mail addresses because I may run for statewide office someday and so or, what I try to do if I have time, is then as soon as I vote on a pro-gun right legislation I send out an e-mail saying I voted for this pro-gun right legislation to my whole group of people that I have categorized because we have categories, we have categories in their contacts so those are the, my second amendment category.

In further questioning, legislator J notes that she categorizes constituents by creating e-mail folders that describe the constituent. For example, she has a "Republican" e-mail folder and e-mail address list, a "Pro-Life" e-mail folder and e-mail address list, and so on. She then has her legislator assistant send messages to these focused e-mail lists as appropriate. In subsequent conversations during the review process, legislator J made it clear that the goal of this behavior is not to surround herself with like-minded people or to filter out constituents who disagree with her, but rather, to simply maintain a database of those who agree with her on issues. Legislator E takes a similar approach:

> I will save an e-mail that I'm going to take action on; if somebody asks me to do something, I hang on to that until it's done. Other than that, the only e-mails that I'll save will be e-mails from people who support me strongly on certain issues, be it guns or what have you. And I simply save that so that I can use the e-mail addresses later on for campaigning or fundraising purposes.

In a third example, legislator I who also maintains e-mail lists of like-minded constituents explains his criteria for adding constituents to his mailing list:

Well, if somebody—, if you can tell that you're aligned, ideologically, you know any efforts that you do will be efficient because you're going to be working together. It's a waste of time to try to bring on somebody that doesn't agree with you or think the same way you do, or who's fighting you. You're much better off trying to find somebody else who does, rather than convert the person that doesn't. It's a waste of time.

And how do you—

I just let them—, basically, it ends up agreeing to disagree is where you end up. Ninety-nine percent of the time. People don't change. But I have had literally people hug me and say they really like me and even though we've disagreed literally they've . . . [answers phone] So, no, I'm trying to build a movement, and you do that by bringing on people that agree with you. You don't do it by trying to change people that don't. And really the key is coordination. You get things done when everybody wants to do the same thing. The hard thing is herding cats. And when everyone is trying to do different things, you can't do anything. So, I try to bring on people that-, I try to get as close to people as I can that we all share the same values and principles.

Legislator I raises a number of issues. First, that he is trying to build a movement and the only way to build a movement is by "bringing on" people who agree with him. Second, legislator I indicates that movements cannot be built by trying to change constituents who disagree with his value positions. Third, legislator I suggests that engaging individuals who disagree with him ends up in a stalemate, with discussants "agreeing to disagree" and that "people don't change." Legislator I indicates that including individuals in his "movement" which disagree with his values and principles would be like "herding cats." This legislator seems to be expressing the opinion that including constituents who disagree with his values and principles takes the focus out of the "movement" he is trying to build, or at least, makes progress more difficult.

In all of these threads, e-mail is the communication technology that legislators use to filter (by deleting or having their legislator assistant delete) communications from constituents who *disagree* with their policy agenda, and to build communication lists that allow focused communications with *like-minded* individuals. Effectively, e-mail is being used as a way for a legislator build databases of like-minded constituents. In another example of this behavior, legislator E indicates that he deletes (or has his legislator assistant delete) bulk e-mails from individuals who disagree with his policy position but saves and responds to bulk e-mails that agree with his policy position:

And if the bulk e-mails are against my position, I just delete, delete, delete, delete. And it's nice that you can group them title—they usually all have the same title and you can do mass deletions. Yeah, I mean, I'll read the first one,

> but after that you pretty much know what's going on. If it's an e-mail that . . . the position I'm supporting, I'll usually . . . either I'll give a very quick, cryptic yes thank-you, you know, I agree, I support you, or I'll do a little one paragraph reply. I'll give it to my assistant and I'll just tell her to cut and paste those into any and all of these e-mails that come in.

Text Messaging

Second pass themes for text messaging communications include importance, lobbyists, age, trust, texting on the floor, and time savings. Text messaging behavior is grounded in 786 quotes and 44 individual codes. As a communication technology used by legislators, text messaging arose somewhat as a surprise. There were indications that text messaging was useful to legislators in the quantitative survey; two legislators wrote in text messaging in the "other communication technologies used" category.

Lobbyists

Legislators mentioned the use of text messaging to communicate with lobbyists. Importantly, communication with lobbyists is related to the use of text messaging on the floor of the House and Senate. Legislator K, providing a link between text messaging and lobbyists:

> Texting I will do a little bit of, not a lot, with a, again, mostly with just family members, personal kind of stuff. Occasionally there are—, I mean, I have the text address for all of the other members on my phone, but I don't text with any of them very often. But occasionally I would. There are like I say probably three or four lobbyists who I would text with. But, again, those are sort of—, they're typically not-, they're lobbyists who run organizations. They are not the typical paid lobbyists.

Texting on the Floor

The use of text messaging while legislators are engaged in floor debates in the house and senate is one of the most interesting uses of communication technology discovered during this dissertation research. In effect, lobbyists are feeding information to legislators during floor debates, and legislators are using that information "real time" to bolster their argument. In addition, floor rules intended to stop legislators from communicating (rules such as remaining in their seats during the third and final reading of a bill and formally recognizing legislators prior to their entry into the discussion on the floor) are being circumvented. Here is legislator B:

> I will give you the prime example and an outsider would not have a clue, and this one is sort of interesting from a couple points of view, because all the communication that is done by our computers, although we are all interlinked, and we can do a communication to all the members of our caucus at the same time, all the members of the legislature, when we are on the floor and we are in debate and in discussion, we tend to communicate among ourselves through our private devices, not our public.

Like texting from a phone to a phone?

> Yes. And I have observed some of my colleagues being prompted to engage in debate on the floor through their private devices from lobbyists in the gallery who are sitting, observing the dialogue, they don't have a direct voice into the discussion but they text, I have literally seen colleagues read arguments from their phone.
>
> What an awesome use of technology, but what a challenge for, in my case, it happened on a bill that I knew my counter arguer, I don't know if that's a word, was not particularly knowledgeable about, that individual simply read from things that were given to them, as argument point counter point and so I raised an issue I knew full well the individual was not informed about and was incapable of answering based on their understanding of what the legislation did, only to watch him read the answer from the outside source who was paid to be there to argue. A paid pawn in the discussion who ethically has no place in the discussion at that point.

Legislator E, indicating that such texting is rare:

> Yeah . . . mostly for personal stuff; I very rarely . . . . Sometimes on the floor you'll get texted by another member during debate or something, but it's pretty rare. . . . not much texting

And legislator K, suggesting that the use of texting on the floor is not so rare:

> I've no doubt that it is a common phenomenon. It's just not one that I personally have much to do with. And again this is not a criticism of anybody else. In fact, to some degree I'm probably jealous of how adept they are at that process. But that would not be me. When I was in the House, and on third read of bills in the House, you know, you can't leave your chair. Once you're in the third read you have to stay there. Now, you can e-mail somebody or you can text another member, you know, across the chamber, you know: How are you going to vote on this one? Something like that.

This section closes with legislator L's perspective on texting on the floor:

> On my phone checking e-mails all the time. Text seems to be the upcoming thing that everybody wants to talk to you, even on the floor when we go in session there would be a lot of texting going on to talk about pros and cons of each piece of legislation. Rather than talking on the phone, it becomes, you know, "Did you know this was in the bill? How are we gonna fix this?" or "What—, can this be done?" that kind of stuff, so—For example a question may be asked on the floor about a certain part of a bill. And a lobbyist would say, "Here's the answer to the question that you're asking." So it—

The implications of the real-time use of text messaging during legislative floor debates could be dedicated to an entire book and will not be explored here; however, they can be summed up by one question: What are the democratic and institutional implications of unelected officials engaging in floor debates as legislators parrot questions and responses generated by lobbyists and special interest groups? This is a topic for future research.

Social Media

Second pass themes for face-to-face communications include YouTube™, Facebook™, and Twitter™. Social media behavior is grounded in 328 quotes and 41 individual codes. In general, the Arizona legislators interviewed do not use social media, and largely indicate that time constraints are a significant reason why. Here is legislator D on time constraints:

> No, I don't do any social media. I have enough time keeping up with my legislative e-mail account, my personal e-mail account and my REDACTED—I'm a REDACTED—e-mail account. The thought of having to respond to people who are on Twitter™, Facebook™, LinkedIn and all the rest is just too daunting.

Legislator B, suggesting that more ambitious legislators use social media:

> I have watched my colleagues who are ambitious who have a political future in mind being much more active in Facebook™ and Twitter™.

Several legislators interviewed commented that they knew they should be using social media, but somehow, they never get around to it. Examination of the quantitative survey responses offers a slightly contradictory finding: with twenty-nine out of fifty-seven legislators responding that they never use Twitter™ to communicate with constituents, thirty legislators indicating they never use Twitter™ to communicate with peers, sixteen legislators indicating they never use Facebook™ to communicate with constituents, and twenty-one legislators indicating they never use Facebook™ to communicate with peers. So, while some legislators are using social media, legislators who agreed to be interviewed tended not to use it.

QUANTITATIVE ANALYSES

Demographic, Political, and Institutional Effects

Peer Communications

Legislators were asked to rate the frequency of use of, and the importance they assigned to, Internet-enabled communications with their peers (other legislators). Importance was defined for legislators as "Importance is related to the likelihood that you will respond favorably to a request received from a peer (or constituent) via one of the communication technologies shown in table 4.1." Frequency of use categories were "Do Not Use," "Use Annually," "Use Monthly," "Use Weekly," "Use Daily," and "Use Hourly." Importance categories were "Do not use," "Not Important," "Slightly Important," "Moderately Important," "Important," and "Very Important." Both frequency of use and importance categories were coded 0 through 5 where 0 was do not use and 5 represented the highest frequency of use and importance. Table 4.1 highlights the frequency of use and importance of legislator use of Internet-enabled communication technologies when communicating with their peers.

Referencing table 4.1 when communicating with other legislators, legislators most frequently use the e-mail, followed by blogs (more on this unexpected result later) and then YouTube™, Facebook™, Twitter™, web pages, and text messages. Legislators find e-mail most important when communicating with their peers, followed by blogs, Facebook™, YouTube™, Twitter™, web pages, and text messages.

Inferential analyses start with bivariate ordinal regression models to examine the relationships between frequency of use (dependent variable) and importance (dependent variable) for each form of Internet-enabled communication technology when communicating with peers, and the various DPI variables that form the independent variables for each ordinal bivariate regression. These variables include age, years in office, chamber (Senate or House) gender, education, race, nine-category political party (a measure of conservatism from strongly progressive (coded as 0) and strongly conservative (coded as 8), gerrymandering measures, majority/minority party status, polarization (binary and ordinal forms), professional/citizen legislature (binary and ordinal forms), term limits, conflict scale, and Trustee/Delegate scale.

Table 4.1 Frequency of Use and Importance Peer Descriptive Statistics

Peer Frequency of Use	*e-mail*	*Twitter™*	*Facebook™*	*YouTube™*	*Web Pages*	*Blogs*	*Text Messages*
Do Not Use	10	945	529	686	1,206	181	1,190
Use Annually	17	63	75	190	87	23	146
Use Monthly	87	166	244	300	122	118	151
Use Weekly	299	203	324	197	109	354	80
Use Daily	868	195	380	192	58	633	33
Use Hourly	374	39	74	48	16	327	7
Mean/Mode	3.885/4	1.23/0	2.03/0	2.11/0	0.608/0	3.354/4	0.532/0
σ	0.891	1.619	1.698	1.539	1.199	1.479	1.034
Importance							
Do Not Use	7	870	560	730	1,018	156	1,080
Not Important	12	248	263	302	259	31	281
Slightly Important	63	237	349	280	160	116	145
Moderately Important	213	129	221	163	91	211	53
Important	590	70	145	79	32	491	23
Very Important	733	37	67	43	22	595	10
Mean/Mode	4.204/5	0.989/0	1.502/0	1.178/0	0.689/0	3.647/5	0.548/0
σ	0.914	1.338	1.373	1.373	1.132	2.405	0.964

Self-Generated by Author using Microsoft Word.

Internet-Enabled Communication Technology Frequency of Use with Peers

Bivariate ordinal logistic regressions were run for the frequency of use of Internet-enabled communications when communicating with peers, and DPI variables. The results of these frequency of use ordinal bivariate regressions are summarized in the sections below.

One of the first things to note about the bivariate regressions in this chapter is that the amount of variation in the dependent variable (in this case, peer frequency of use and DPI variables) by variations in the independent variables, pseudo r-squared is very small. The largest pseudo r-squared in the bivariate regression models is 1.86 percent (frequency of using Facebook™ and age), meaning that only 1.86 percent of the variation in the frequency of use (in this case) of using Facebook™ is explained by the age of the legislator. All other pseudo r-squared values were less, and many were as low as 0.1 percent. Put succinctly, while there are statistically significant relationships between peer communication technology frequency of use and DPI variables, they do not explain much of the variation in peer communication technology frequency of use. Later in this chapter, full regression models will be presented which examine combinations of variables; r-squared values typically increase and will be reported in the full regression models.

Text Message Communications

With respect to peer frequency of use, legislator age and being black were statistically significant. A one-year increase in age is associated with a 1.5 percent decrease in the probability that a legislator will indicate that text messages are used hourly (over all other frequency of use categories) to communicate with their peers, ceteris paribus. A much more significant relationship was uncovered between the peer frequency of use of text messages and black legislators. Compared to white legislators, black legislators are 125 percent more likely to indicate that they use text messages hourly (over all other frequency of use categories) to communicate with other legislators, ceteris paribus.

E-mail Communications

The relationships between the use of e-mail to communicate with peers and demographics were both related to legislator polarization. Both the polarization binary variable (legislators indicating strongly conservative or strongly progressive political party affiliation were coded as 1, all other political party affiliations were coded as 0) and the ordinal polarization variable were statistically significantly related to peer frequency of use. With respect to polarization, a one unit increase in indicated political party polarization was associated with a 14.8 percent increase in the probability that a legislator would indicate hourly use (over all other frequency of use categories) of e-mail to communicate with their peers, on average, all else equal. With respect to the binary polarization variable, polarized legislators were 38.6 percent more likely to choose "Use Hourly" (over all other frequency of use categories) to communicate with their peers via e-mail than their less polarized peers, ceteris paribus.

Twitter™ Communications

Given President Trump's use of Twitter™ as this book is being written, association of the statistically significant DPI variables with Twitter™ use may be slightly more interesting than that with other communication technologies. The political, demographic, and institutional variables that are statistically significantly correlated with Twitter™ use include age, years in office, political ideology (*partyid*), polarization, gender, race, gerrymandering, and professional legislature status. For clarity, two of these variables will be interpreted for the reader. With respect to the ordinal polarization variable, a one unit increase in polarization is associated with a 13.6 percent ($z \leq 0.01$) increase in the probability that a legislator will indicate hourly use of Twitter™ over all other frequency of use categories, ceteris paribus. With respect to race, compared to white legislators, black legislators are 147.5 percent ($z \leq 0.001$) more likely to choose hourly use of Twitter™ over all other frequency of use categories, on average, all else equal.

Facebook™ Communications

The statistically significant relationships between Facebook™ use to communicate with peers and the DPI variables examined include age, years in office, Senators (compared to Representatives), polarization (ordinal and binary), gender, race, gerrymandering (comprehensive, Polsby-Popper Schwartzberg, and Convex Hull), and the Trustee/Delegate scale. As in the previous section, two relationships will be examined. Compared to Representatives, Senators are 24.4 percent ($z \leq 0.01$) less likely to choose hourly use of Facebook™ over all other frequency of use categories to communicate with their peers, on average, all else equal. Compared to female legislators, male legislators are 28.2 percent ($z \leq 0.01$) less likely to choose "Use Hourly" over all other frequency of use categories when using Facebook™ to communicate with their peers, on average, all else equal.

YouTube™ Communications

With respect to YouTube™ communications with peers, political ideology, race, gerrymandering, and legislator perception of policy conflict in their district have statistically significant relationships. As with all other statistically significant relationships between peer use of communication technology and race, black legislators are more likely to use YouTube™ to communicate with their peers than their white counterparts. Black legislators are 87.9 percent ($z \leq 0.01$) more likely to indicate hourly use of YouTube™ over all other frequency of use categories than white legislators, ceteris paribus. A one unit increase in conservativism (measured as political party) is associated with a 3.1 percent ($z \leq 0.05$) decrease in the probability that a legislator will indicate hourly use of YouTube™ over all other frequency of use categories to communicate with their peers, on average, all else equal.

Web Page Communications

There were only two statistically significant relationships between the use of web pages to communicate with peers: age and legislator Delegate behavior. With respect to age, a one-year increase in legislator age is associated with a 1.6 percent ($z \leq 0.01$) decrease in the probability that a legislator will indicate hourly use of web pages to communicate with their peers over all other frequency of use categories, ceteris paribus. A 1 percent increase in a legislator acting like a Delegate is associated with 0.3 percent increase in the probability that a legislator will indicate hourly use of web pages to communicate with their peers over all other frequency of use categories, ceteris paribus.

Blog Communications

Blog communications by legislators can be hosted by the state legislature blog sites or can be a private blog site hosted by the legislator independent of the state's websites. Based on existing research on legislator use of blogs, legislators use blogs primarily to communicate with constituents during campaigning (Hamajoda, 2016; Helfert, 2017); it is somewhat of a surprise to see the significance that legislators place on blogs. It is difficult not to wonder if legislators failed to understand the question regarding the frequency of use and importance of blogs, yet, the relationships uncovered were spread across several questions, suggesting that legislators both use and value blogs for more than just campaigning. The statistically significant variables associated with using blogs to communicate with peers include age, Senators (compared to Representatives), gerrymandering (all measures of gerrymandering except the Convex Hull measure), minority party status, professional legislator status (two measures), term limits, and the Trustee/Delegate scale. With respect to professional legislature status, legislators from states with professional legislatures are 42.6 percent ($z \leq 0.01$) more likely to indicate hourly use of blogs (over all other frequency of use categories) to communicate with peers than citizen legislators, on average, all else equal. Legislators who were in the minority party in one chamber were 55.4 percent ($z \leq 0.01$) more likely to indicate hourly use of blogs (over all other frequency of use categories) to communicate with peers than citizen legislators, on average, all else equal.

Peer Communication Technology Frequency of Use Models

Three models were created to determine the relationships between peer frequency of use of Internet-enabled communication technology and DPI variables. The basic model contains legislator demographic information, the intermediate model adds primary political and institutional variables, and the advanced model adds secondary institutional variables. Additionally, the results of these three models are not presented in a table, but rather, are discussed in the paragraphs following the models presented below. These models use a dependent variable *pfreqiect* which is the average frequency of use of all seven Internet-enabled communication technologies when used to communicate with peers. In effect, these models look at the overall frequency of use for Internet-enabled communication technology. The basic model contains legislator demographic information, the intermediate model adds primary political and institutional variables, and the advanced model adds secondary institutional variables. Additionally, the results of these three models are not presented in a table, but rather, are discussed in the paragraphs following the models presented below:

Basic Model: $pfreqiect = \beta_0 + \beta_1(age) + \beta_2(race) + \beta_3(male) + \beta_4(education) + \mathcal{E}$.

The basic model is statistically significant for age (odds ratio = 0.963, $z \leq 0.001$), legislators in the "other" race category compared to white legislators (odds ratio = 0.659, $z \leq 0.05$), black legislators compared to white legislators (odds ratio = 1.776, $z \leq 0.05$), and gender (odds ratio = 0.694, $z \leq 0.001$). Interpreting a single variable in the basic model for clarity, compared to female legislators, male legislators are 30.4 percent less likely to choose "Use Hourly" for the use of Internet-enabled communications with peers over all other communication technology frequency of use categories, ceteris paribus.

Intermediate Model: $pfreqiect = \beta_0 + \beta_1(age) + \beta_2(race) + \beta_3(male) + \beta_4(education) + \beta_5(conservativism) + \beta_6(minority\ status) + \beta_7(years\ in\ office) + \beta_8(senate) + \mathcal{E}$.

The intermediate model is statistically significant for age (odds ratio = 0.966, $z \leq 0.001$), black legislators compared to white legislators (odds ratio = 0.651, $z \leq 0.01$), gender (odds ratio = 0.732, $z \leq 0.001$), political ideological conservativism (odds ratio = 0.961, $z \leq 0.05$), minority party status in both chambers and the governorship (odds ratio = 0.780, $z \leq 0.05$), and years in office (odds ratio = 0.985, $z \leq 0.05$). Interpreting a single variable in the intermediate model for clarity, a one-year increase in legislator years in office is associated with a 1.5 percent decrease in the likelihood a legislator will select "Use Hourly" for the use of Internet-enabled communications with peers over all other communication technology frequency of use categories, ceteris paribus.

Full Model: $pfreqiect = \beta_0 + \beta_1(age) + \beta_2(race) + \beta_3(male) + \beta_4(education) + \beta_5(conservativism) + \beta_6(minority\ status) + \beta_7(years\ in\ office) + \beta_8(senate) + \beta_9(gerrymandering) + \beta_{10}(professional\ legislature\ binary) + \beta_{11}(term\ limits) + \beta_{12}(conflict\ scale) + \beta_{13}(Trustee/Delegate\ scale) + \beta_{14}(polarization\ binary) + \mathcal{E}$.

The full model is statistically significant for age (odds ratio = 0.967, $z \leq 0.001$), "other" race legislators compared to white legislators (odds ratio = 0.653, $z \leq 0.05$), gender (odds ratio = 0.724, $z \leq 0.01$), political ideological conservativism (odds ratio = 0.958, $z \leq 0.05$), gerrymandering (odds ratio = 1.004, $z \leq 0.05$), and polarization (odds ratio = 1.359, $z \leq 0.05$). Interpreting a single variable in the full model for clarity, a legislator who self-identified as strongly progressive or strongly conservative in office is associated with a 35.9 percent increase in the likelihood a legislator will select "Use Hourly" for the use of Internet-enabled communications with peers over all other communication technology frequency of use categories, ceteris paribus.

The previous section examines frequency of use models using an average of all seven Internet-enabled communication technologies. Table 4.2 contains the results of the full regression model using *each* individual Internet-enabled communication technology as the dependent variable, and all of the DPI variables available. The results of these full ordinal regression models are shown in table 4.2.

A review of the age variable in table 4.2 shows that for all statistically significant Internet-enabled communication technologies, the use of the technology to communicate with other legislators decreases as a function of legislator age. This finding is consistent with much of the research on non-legislators covering the use of computers in general and in Internet-related communication technology as a function of age, which show that the use of computers and Internet-related communication technology decreases with increasing age (Juznic et al., 2006; Knight and Pearson, 2005; Schleife, 2006; Thayer and Ray, 2006). Another interesting general relationship highlighted

Table 4.2 Peer Frequency of Use Full Regression Models, Odds Ratio

Peer Frequency of Use	*Text*	*e-mail*	*Twitter™*	*Facebook™*	*YouTube™*	*Web Pages*	*Blogs*
Age	0.984*	0.995	0.949***	0.970***	1.000	0.986*	0.968***
Years in Office	1.001	0.998	0.994	0.962***	0.986	1.002	1.013
Senate (Compared to House)	1.154	0.926	1.105	0.825	0.919	0.939	1.303*
Political Party (progressive to conservative)	0.980	0.968	0.903***	0.984	0.973	0.967	1.016
Polarization Binary	1.476*	1.415*	1.474*	1.373*	1.059	1.054	1.123
Male	0.884	0.979	0.625***	0.719*	0.315	1.048	0.934
Education	1.000	1.007	1.047*	0.989	1.023	1.064*	0.984
Race (Compared to White)							
Black	1.895*	1.100	1.256	1.292	1.776*	1.184	1.285
Other	1.104	0.829	0.668*	0.694*	0.886	0.905	0.739
Gerrymandering (Comprehensive)	1.002	1.005*	1.000	0.998	1.005*	1.002	1.008
Minority One Chamber	0.721	1.210	0.821	1.070	0.985	0.729	1.706**
Minority Two Chambers	0.788	0.962	0.750	0.885	0.978	1.038	1.131
Minority Two Chambers and Governorship	0.982	0.889	0.790	0.774	0.893	0.085	0.989
Professional Legislature Binary	1.088	1.125	1.497*	0.991	1.093	0.780	1.441*
Term Limits	0.823	1.082	0.740*	1.093	0.954	1.069	1.012
Conflict Scale	1.004	0.998	1.006	1.007*	1.008*	1.010*	1.007*
Trustee/Delegate Scale	1.000	1.001	1.000	1.003	1.002	1.004*	1.002

* z ≤.05, ** z ≤.01, *** z ≤.001.
Self-Generated by Author using Microsoft Word.

in table 4.2 is that every statistically significant finding associated with legislator polarization and Internet-enabled communication technology shows a substantial increase in polarization. These polarization levels range from 37.3 percent to 47.6 percent increases in polarization as the peer frequency of use of that communication technology increases. Another significant finding is that an increase in the use of Internet-enabled communication technology is associated with *increase* in the perception of constituent policy conflict in their district.

Less frequent but still consistent statistically significant findings are that male legislators use Twitter™ and Facebook™ to communicate with their peers less than female legislators and that black legislators use text messages and YouTube™ more than their white peers. In one final consistent result, a decrease in district gerrymandering is associated with increases in e-mail and YouTube™ use. With these general themes presented, let's examine each individual Internet-enabled communication technology.

Text Message Communications

In the full regression model for text messages, age, polarization, and race were statistically significant. With respect to polarization, legislators who identified as strongly progressive or strongly conservative were 47.6 percent ($z \leq 0.05$) more likely to choose "Use Hourly" for communicating with their peers via text messages over all other communication technology frequency of use categories, than their more moderate peers, on average, all else equal. Compared to white legislators, black legislators were 89.5 percent ($z \leq 0.05$) more likely to choose "Use Hourly" for communicating with their peers via text messages, over all other communication technology frequency of use categories, than their more moderate peers, on average, all else equal.

E-mail Communications

With respect to the full regression model for e-mail, only polarization and gerrymandering (comprehensive), were statistically significant. Legislators who identified as strongly conservative or strongly progressive were 41.5 percent ($z \leq 0.05$) more likely to choose "Use Hourly" when e-mailing their peers, on average, all else equal. With respect to the comprehensive gerrymandering variable, a one unit increase in gerrymandering score (varies from 110 to 262), meaning a *decrease* in gerrymandering, is associated with a 0.5 percent ($z \leq 0.05$) increase in the use of e-mail to communicate with peers. This relationship seems intuitive since gerrymandered districts are "safer" (Chen and Cottrell, 2016; Gardner, 2006; Klain, 2007; Litton, 2012) from a reelection perspective, and if indeed legislators are "single minded seekers of reelection" (Mayhew, 2004, 5), and if legislators communicate primarily to be reelected (as this book maintains), then legislators in less competitive districts can be expected to communicate less.

Twitter™ Communications

The Twitter™ model has statistically significant relationships with age, conservativism, polarization, gender, education, the "other" race category, professional legislature status, and term limits. With respect to term limits, legislators in states with term limits are 26 percent less likely to choose "Use Hourly" to communicate with their peers using Twitter™ than legislators from states without term limits, on average, all else equal. With respect to ideological conservativism, a one unit increase in legislator conservativism is associated with a 2.7 decrease in the probability that a legislator will select "Use Hourly" for Twitter™ communications with their peers, on average, all else equal.

Facebook™ Communications

With respect to Facebook™ use with peers, statistically significant relationships included age, years in office, polarization, gender, the "other" race category (non-black races), and the conflict scale. A one-year increase in a legislator's time in office is associated with a 3.8 percent decrease in the probability that a legislator will select "Use Hourly" for Facebook™ communications with other legislators, on average, all else equal. Compared to whites, non-black races are 30.6 percent less likely to choose "Use Hourly" for Facebook™ to communications with other legislators, ceteris paribus.

YouTube™ Communications

YouTube™ communications with peers is associated with three statistically significant variables: blacks, gerrymandering, and the conflict scale. Blacks are 77.6 percent more likely to choose "Use Hourly" (over all other frequency of use categories) to communicate with their peers via YouTube™ than white legislators, on average, all else equal. With respect to the conflict scale finding, a 1 percent increase in a legislator's perception of constituent policy conflicts in their district is associated with a 0.8 percent increase in the probability that a legislator will choose "Use Hourly" (over all other frequency of use categories) to communicate with their peers via YouTube™, ceteris paribus.

Web Page Communications

Legislator use of web pages to communicate with their peers is associated with four statistically significant independent variables: age, education, a legislator's perception of constituent policy conflict in their district, and a legislator's behavior as a Delegate. With respect to education, a one-year increase in legislator education is associated with a 6.4 percent increase in the probability that a legislator will choose "Use Hourly" for web page communication with their peers, ceteris paribus. With respect to legislator age, a one-year increase in legislator age is associated with a 1.4 percent increase in the probability that a legislator will choose "Use Hourly" (over all other frequency of use categories) for web page communications with their peers, on average, all else equal.

Blog Communications

Legislator use of blogs to communicate with other legislators is associated with age, Senate membership (compared to House membership), minority party status in one legislative chamber, professional legislature status, and a legislator's perception of constituent policy conflict in their district. Senators are 30.3 percent more likely to select "Use Hourly" (over all other frequency of use categories) for communications with their peers using blogs than their Representative counterparts, on average, all else equal. Legislators who are in the minority party in one legislative chamber in their state are 70.6 percent more likely to select "Use Hourly" (over all other frequency of use categories) to communicate with their peers using blogs than their majority counterparts, on average, all else equal.

Internet-Enabled Communication Technology Importance, Peers

Bivariate ordinal logistic regressions were run for the importance of Internet-enabled communications when communicating with peers, and DPI variables. The results of these importance ordinal bivariate regressions are discussed in the sections below.

A review of the age variable suggests that for all statistically significant Internet-enabled communication technologies, the importance of the technology to communicate with other legislators decreases as a function of legislator age. This finding is consistent with the frequency of use regressions for age shown in table 4.2 which highlight that the use of Internet-enabled communication decreases with increasing legislator age. Although no empirical data or analyses could be located for the relative importance of Internet-enabled communication to legislators, it is reasonable to expect that the decrease in the use of Internet-enabled communication technologies shown in table 4.2 would correlate with a decrease in the importance of Internet-enabled communication technology as a function of age. Spearman's coefficient was used to examine the correlation between frequency of use and importance for Internet-enabled communication technology: Spearman's rho = 0.851, $p \leq 0.001$ suggesting that there is a strong statistically significant connection between the frequency of use and importance of Internet-enabled communication technology when used to communicate with peers and constituents. The relationship between the frequency of use and importance of Internet-enabled communication technology when used with constituents is stronger (Spearman's rho = 0.851, $p \leq 0.001$) than when Internet-enabled communication technology is used to communicate with peers (Spearman's rho = 0.741, $p \leq 0.001$).

Another interesting general bivariate relationship is that every statistically significant finding associated with legislator years in office shows that legislators find Internet-enabled communication technologies less important the longer they are in office. For all statistically significant bivariate regression results, male legislators find Internet-enabled communication technologies less important than female legislators, and black legislators find Internet-enabled communication technologies more important than white legislators.

Less frequent but still consistent statistically significant findings are that legislators who identify as strongly progressive or strongly conservative find Internet-enabled communication technology more important than their more moderate peers. Additionally, in every statistically significant case, professional legislators find Internet-enabled communication technology more important than their citizen legislator equivalents. With these general observations presented, let's examine each individual Internet-enabled communication technology.

Text Message Communications

The importance of text messages when used to communicate with peers is statistically significantly associated with legislator age, years in office, ideological conservativism, polarization, gender, race (black and "other" races), and term limits. Legislator age, years in office, conservativism, and gender (male) are all associated with decreases in the importance of text messages to communicate with peers, while polarization, black, and other races compared to white legislators and term limits are associated with increases in the importance of using text messages to communicate with other legislators. As in other sections, two of these results will be fully evaluated and the rest will be left as an exercise for the reader. With respect to term limits, legislators in states with term limits are 32.3 percent more likely to select "Very Important" for the importance of text message communications with their peers than all other text message importance categories, on average, all else equal. Compared to female legislators, male legislators are 30.2 percent less likely to choose "Very Important" for text message communications with other legislators over all other text message importance categories, on average, all else equal.

E-mail Communications

E-mail importance when communicating with other legislators is statistically significantly associated with legislator years in office, polarization (both measures), gender, race (other), professional legislature status, term limits, and legislator Delegate behavior. Increases in the importance of e-mail communications with peers is associated with legislator polarization variables, professional legislature status, and legislator Delegate behaviors. Decreases in the importance of e-mail when communicating with peers is associated with legislator years in office, gender (male), race (other), and term limits. Legislators in states with term limits are 18.9 percent less likely to choose "Very Important" for the importance of e-mail communications with other legislators over all other e-mail importance categories, ceteris paribus. With respect to gender, compared to female legislators, male legislators are 35.9 percent less likely to choose "Very Important" for the importance of e-mail communications with other legislators over all other e-mail importance categories, on average, all else equal.

Twitter™ Communications

The importance of Twitter™ communications is statistically significantly correlated with age, years in office, Senate membership, gender, race (black), gerrymandering (Convex Hull measure), and both professional legislature measures. Increases in the importance of Twitter™ use with peers are associated with polarization, race, and both professional legislature measures. Decreases in the importance of using Twitter™ to communicate with peers are associated with legislator age, years in office, political ideological conservativism, gender, and gerrymandering. A one unit increase in gerrymandering (decreased gerrymandering) is associated with a 1.4 percent decrease in the probability that a legislator will choose "Very Important" for Twitter™ communications with peers over all other Twitter™ importance categories, on average, all else equal.

Facebook™ Communications

The statistically significant relationships between the importance of Facebook™ use when communicating with peers include legislator age, years in office, Senate membership, gender, race (black compared to white), gerrymandering, legislator perception of policy conflict in their district, and legislator Delegate behavior. Increases in the importance of Facebook™ communications with other legislators are associated with race, conflict scale, and Delegate scale. Decreases in the importance of Facebook™ communications when communicating with other legislators are associated with legislator age, legislator years in office, Senate membership, gender (being male), and gerrymandering (Convex Hull). A 1 percent increase in legislator Delegate behavior is associated with a 0.3 percent increase in the probability that a legislator will choose "Very Important" for Facebook™ communications with other legislators over all other Facebook™ importance categories, on average, all else equal. A one-year increase in legislator age is associated with a 3.5 percent increase in the probability that a legislator will select "Very Important" for Facebook™ communications with their peers over all other Facebook™ importance categories, ceteris paribus.

YouTube™ Communications

With respect to the importance of YouTube™ communications with peers, statistically significant relationships are correlated with legislator age, legislator years in office, legislator polarization, race (black compared with white), gerrymandering (Polsby-Popper measure), and the Delegate scale. Increases in the importance of YouTube™ communications when communicating with peers are associated with polarization, race (black compared with white), the Polsby-Popper gerrymandering measure, and the legislator Delegate scale. Decreases in the importance of YouTube™ communications with other legislators are associated with legislator age and legislator years in office. A one-year increase in the time that a legislator has been in office is associated with a 1.5 percent decrease in the probability that a legislator will choose "Very Important" for YouTube™ communications with their peers over all other YouTube™ importance categories, on average, all else equal. With respect to the Delegate scale, a 1 percent increase in legislator behavior as a Delegate is associated with a 0.3 percent increase in the probability that a legislator will choose "Very Important" for YouTube™ communications with their peers over all other YouTube™ importance categories, on average, all else equal.

Web Page Communications

The importance of web page communications with other legislators is statistically significantly correlated with legislator age, years in office, political ideology, gender, race (black and "other" compared to white legislators), minority status in one chamber (compared to majority status), and term limits. Statistically significant positive correlations with the importance of web page communications with peers occur with race (black and other), and term limits, while statistically significant negative correlations occur with legislator age, years in office, political ideology, gender, and minority party status in one legislative chamber. A one unit increase in political ideological conservativism is associated with a 3.7 percent decrease in the probability that a legislator will choose "Very Important" for web page communications with their peers over all other web page importance categories, on average, all else equal. A one-year increase in legislator age is associated with a 0.8 percent decrease in the probability that a legislator will choose "Very Important" for web page communications with their peers over all other web page importance categories, on average, all else equal.

Blog Communications

The importance of blog communications with other legislators is statistically significantly related to legislator age, years in office, Senate membership, both polarization variables, gender, *every* gerrymandering variable except Convex Hull, professional legislature status (both measures), and term limits. The variables associated with increases in the importance of blog communications with peers include Senate membership, both polarization measures, all gerrymandering measures except Convex Hull, both professional legislature measures, and term limits. The variables associated with decreases in the importance of blog communications with peers include age, years in office, and gender. A one-year increase in a legislator's time in office is associated with a 2.5 percent decrease in the probability that a legislator will choose "Very Important" for blog communications with their peers over all other blog importance categories, on average, all else equal. A one unit increase in gerrymandering score (comprehensive measure), meaning less gerrymandering is associated with a 0.5 percent increase in the probability that a legislator will choose "Very Important" for blog communications with their peers over all other blog importance categories, on average, all else equal.

Peer Communication Technology Importance Models

Three models were created to determine the relationships between the importance of Internet-enabled communication technology and DPI variables. These models use a dependent variable *pimportiect* which is the average of all seven Internet-enabled communication technologies when used to communicate with peers. In effect, these models look at the overall importance of Internet-enabled communication technology. The basic model contains legislator demographic information, the intermediate model adds primary political and institutional variables, and the advanced model adds secondary institutional variables. Additionally, the results of these three models are not presented in a table, but rather, are discussed in the paragraphs following the models presented below:

Basic Model: $pimportiect = \beta_0 + \beta_1 (age) + \beta_2 (race) + \beta_3 (male) + \beta_4 (education) + \varepsilon$.

The basic model is statistically significant for age (odds ratio = 0.962, $z \leq 0.001$), black legislators compared to white legislators (odds ratio = 1.755, $z \leq 0.05$), and gender (odds ratio = 0.604, $z \leq 0.001$). Interpreting a single variable in the basic model for clarity, compared to female legislators, male legislators are 39.6 percent less likely to select "Very Important" for the use of Internet-enabled communications with peers over all other communication technology importance categories, ceteris paribus.

Intermediate Model: $pimportiect = \beta_0 + \beta_1 (age) + \beta_2 (race) + \beta_3 (male) + \beta_4 (education) + \beta_5 (conservativism) + \beta_6 (minority\ status) + \beta_7 (years\ in\ office) + \beta_8 (senate) + \varepsilon$.

The intermediate model is statistically significant for age (odds ratio = 0.966, $z \leq 0.001$), gender (odds ratio = 0.626, $z \leq 0.001$), political ideological conservativism (odds ratio = 0.964, $z \leq 0.05$), minority party status in both chambers and the governorship (odds ratio = 0.780, $z \leq 0.05$), and years in office (odds ratio = 0. 977, $z \leq 0.05$). Interpreting a single variable in the intermediate model for clarity, a one unit increase in legislator conservativism is associated with a 3.4 percent decrease in the likelihood a legislator will select "Very Important" for the use of Internet-enabled communications with peers over all other communication technology importance categories, ceteris paribus.

Full Model: $pimportiect = \beta_0 + \beta_1 (age) + \beta_2 (race) + \beta_3 (male) + \beta_4 (education) + \beta_5 (conservativism) + \beta_6 (minority\ status) + \beta_7 (years\ in\ office) + \beta_8 (senate) + \beta_9 (gerrymandering) + \beta_{10} (professional\ legislature\ binary) + \beta_{11} (term\ limits) + \beta_{12} (conflict\ scale) + \beta_{13} (Trustee/Delegate\ scale) + \beta_{14} (polarization\ binary) + \varepsilon$.

The full model is statistically significant for age (odds ratio = 0.967, $z \leq 0.001$), gender (odds ratio = 0.693, $z \leq 0.01$), political ideological conservativism (odds ratio = 0.952, $z \leq 0.05$), years in office (odds ratio = 0.978, $z \leq 0.05$), gerrymandering (odds ratio = 1.004, $z \leq 0.05$), and polarization (odds ratio = 1.577, $z \leq 0.01$). Interpreting a single variable in the full model for clarity, a one unit increase in political ideological conservativism is associated with a 4.9 percent decrease in the likelihood a legislator will select "Very Important" for the use of Internet-enabled communications with peers over all other communication technology importance categories, ceteris paribus.

The previous section examines communication technology importance models using an average of all seven Internet-enabled communication technologies. Table 4.3 contains the results of the full regression model using *each* individual Internet-enabled communication technology as the dependent variable and all available DPI variables available. The results of these full ordinal regression models are shown in table 4.3.

A review of the age variable in table 4.3 shows that for all statistically significant Internet-enabled communication technologies, the importance of the technology to communicate with other legislators decreases as a function of legislator age. This finding is consistent with the frequency of use regressions for age shown in tables 5.3 and 5.4 which highlight that the use and importance of Internet-enabled communication technologies decrease with increasing legislator age. Another interesting general relationship highlighted in table 4.3 is that every statistically significant finding associated with legislator years in office shows that legislators find Internet-enabled communication technologies less important the longer they are in office, *even* after controlling for age. For all statistically significant results shown in table 4.3, male legislators find Internet-enabled communication technologies less important than female legislators. For all statistically significant results shown in table 4.3, legislator polarization was correlated with increases in the importance assigned to Internet-enabled communication technologies.

Text Message Communications

With respect to the importance of text messages when communicating with peers, only age is statistically significantly related to text message communications with peers. A one unit increase in legislator age is associated with a 2.8 percent decrease in the probability that a legislator will choose "Very Important" for text message communications with peers over all other text message importance categories, on average, all else equal.

E-mail Communications

The full model for the importance of e-mail communications with other legislators shown in table 4.3 highlights six variables that arc statistically

Table 4.3 Peer Importance Regression Models, Odds Ratios

Peer Importance	*Text Messages*	*e-mail*	*Twitter™*	*Facebook™*	*YouTube™*	*Web Pages*	*Blogs*
Age	0.972***	1.005	0.953***	0.973***	0.999	0.976***	0.965***
Years in Office	0.995	0.980*		0.970***	0.981*	0.982	0.998
Senate (Compared to House)	0.771	1.311*	0.948	0.652**	0.811	0.812	1.463**
Political Party (progressive to conservative)	0.960	0.960	0.981***	0.960*	0.985	0.965	1.027
Polarization Binary	1.122	2.005***	1.528*	1.419*	1.307	0.879	1.596**
Male	0.858	0.702*	0.617***	0.725*	0.951	0.781	0.625***
Education	0.993	1.004	1.034	0.987	0.993	1.010	0.990
Race (Compared to White)	Omitted Category: White	Omitted Category: White	Omitted Category: White	Omitted Category: White	Omitted Category: White	Omitted Category: White	Omitted Category: White
Black	1.537	0.648	1.016	1.241	1.780*	1.715	0.878
Other	1.258	0.654*	0.982	0.902	1.174	1.376	1.007
Gerrymandering (Comprehensive)	1.002	1.001	1.000	0.999	1.003	1.001	1.007***
Minority One Chamber	0.677	0.998	0.862	1.108	0.964	0.579*	1.457*
Minority Two Chambers	0.792	0.970	0.770	0.758	1.046	0.634*	1.005
Minority Two Chambers and Governorship	0.892	0.849	0.801	0.821	0.919	0.690	1.151
Professional Legislature Binary	1.112	1.111	1.340	0.898	0.889	0.775	1.654***
Term Limits	1.025	0.749*	0.736*	0.996	0.914	1.225	1.029
Conflict Scale	0.997	0.998	1.004	1.008*	1.003	1.002	1.006
Trustee/Delegate Scale	1.003	1.003	1.002	1.002	1.003*	1.001	1.001

* z ≤.05, ** z ≤.01, *** z ≤.001.
Self-Generated by Author using Microsoft Word.

significant: years in office, Senate membership, polarization, gender, the "other" race category, and term limits. Negative associations with the importance of e-mail communications with peers include years in office, gender, the "other" race category, and term limits. Positive associations include Senate membership and polarization. A one-year increase in the time a legislator spends in office is associated with a 2 percent decrease in the probability that a legislator will choose "Very Important" for text message communications with peers over all other text message importance categories, on average, all else equal. Legislators who self-identify as strongly progressive or strongly conservative are twice as likely as their more moderate peers to choose "Very Important" for text message communications with peers over all other text message importance categories, on average, all else equal.

Twitter™ Communications

In the full Twitter™ model, four variables were statistically significantly related to the importance of Twitter™ use with peers: age, political ideology (progressive to conservative), polarization, gender, and term limits. Age, political ideology, gender, and term limits are negatively correlated, while polarization is positively correlated. Legislators who self-identify as strongly progressive or strongly conservative are 52.8 percent more likely as their more moderate peers to choose "Very Important" for Twitter™ communications with peers over all other Twitter™ importance categories, on average, all else equal. Legislators in states with term limits are 26.4 percent more likely to choose "Very Important" for Twitter™ communications with peers over all other Twitter™ importance categories, on average, all else equal.

Facebook™ Communications

The importance of Facebook™ communications with other legislators is positively correlated with legislator polarization, and negatively correlated with legislator age, years in office, Senate membership, political conservativism, and gender. Legislators who self-identify as strongly progressive or strongly conservative are 4 percent more likely as their more moderate peers to choose "Very Important" for Facebook™ communications with peers over all other Facebook™ importance categories, on average, all else equal. Compared to Representatives, Senators are 43.8 percent less likely to choose "Very Important" for Facebook™ communications with peers over all other Facebook™ importance categories, on average, all else equal.

YouTube™ Communications

With respect to the importance of YouTube™ communications when communicating with other legislators, three variables had statistically significant correlations: years in office, race (black compared to white), and the Trustee-Delegate scale. Compared to white legislators, black legislators are 78 percent more likely to choose "Very Important" for YouTube™ communications with peers over all other YouTube™ importance categories, on average, all else equal. A 1 percent increase in a legislator's behavior as a Delegate is associated with a 0.3 percent increase in the probability that a legislator will choose "Very Important" for YouTube™ communications with peers over all other YouTube™ importance categories, on average, all else equal.

Web Page Communications

Three variables were statistically significantly correlated with the importance that legislators assign to web page communications with their peers: age, and minority party status in one and two legislative chambers. With respect to age, a one-year increase in legislator age is associated with a 2.4 percent decrease in the probability that a legislator will choose "Very Important" for web page communications with peers over all other web page importance categories, on average, all else equal. Compared to legislators in the majority party, legislators who are in the minority party in two chambers are 36.6 percent less likely to choose "Very Important" for web page communications with peers over all other web page importance categories, on average, all else equal.

Blog Communications

Seven variables had statistically significant relationships with the importance that legislators assign to blog communications with their peers: age, Senate membership, polarization, gender, gerrymandering, minority party status in one chamber, and professional legislature status. Age and gender are negatively correlated with the importance legislators assign to blog communications with their peers, while Senate membership, polarization, gerrymandering, minority status in one chamber, and professional legislature status were positively correlated with the importance legislators assign to blog communications with their peers. Compared to Representatives, Senators are 46.5 percent more likely to choose "Very Important" for blog communications with peers over all other blog importance categories, on average, all else equal. Professional legislators are 65.4 percent more likely to choose "Very Important" for blog communications with peers over all other blog importance categories, on average, all else equal.

CONSTITUENT COMMUNICATIONS

Legislators were asked to rate the frequency of use of and the importance they assigned to Internet-enabled communications when communicating with their constituents. Importance was defined for legislators as "Importance is related to the likelihood that you will respond favorably to a request received from a constituent via one of the communication technologies shown in table 4.4." Frequency of use categories were "Do Not Use," "Use Annually," "Use Monthly," "Use Weekly," "Use Daily," and "Use Hourly." Importance categories were "Do not use," "Not Important," "Slightly Important," "Moderately Important," "Important," and "Very Important." Both frequency of use and importance categories were coded 0 through 5 where 0 was do not use and 5 represented the highest frequency of use and importance. Table 4.4 highlights the frequency of use and importance of legislator use of Internet-enabled communication technologies when communicating with their constituents.

Referencing table 4.4 when communicating with constituents, legislators most frequently use the e-mail, followed by blogs (more on this unexpected result later) and then YouTube™, Facebook™, Twitter™, web pages, and text messages. Legislators find e-mail most important when communicating with their peers, followed by blogs, Facebook™, YouTube™, Twitter™, web pages, and text messages.

Inferential analyses start with bivariate ordinal regression models to examine the relationships between frequency of use (dependent variable) and importance (dependent variable) for each form of Internet-enabled communication technology when communicating with constituents, and the various DPI variables that form the independent variables for each ordinal bivariate regression. These variables include age, years in office, chamber (Senate or House) gender, education, race, nine-category political party (a measure of conservatism from strongly progressive (coded as 0) and strongly conservative (coded as 8), gerrymandering measures, majority/minority party status, polarization (binary and ordinal forms), professional/citizen legislature (binary and ordinal forms), term limits, conflict scale, and Trustee/Delegate scale.

Internet-Enabled Communication Technology Frequency of Use with Constituents

Bivariate ordinal logistic regressions were run for the frequency of use of Internet-enabled communications when communicating with constituents, and DPI variables. The results of these frequency of use ordinal bivariate regressions are summarized in in the sections below.

Table 4.4 Frequency of Use and Importance Constituent Descriptive Statistics

Constituent *Frequency of Use*	*e-mail*	*Twitter™*	*Facebook™*	*YouTube™*	*Web Pages*	*Blogs*	*Text Messages*
Do Not Use	933	469	651	1,236	461	1,225	221
Use Annually	64	69	203	81	55	142	332
Use Monthly	143	216	331	111	198	139	601
Use Weekly	224	381	258	104	318	61	361
Use Daily	193	389	125	42	410	21	61
Use Hourly	40	75	23	9	153	4	8
Mean/Mode	1.249/0	2.236/0	1.417/0	0.523/0	2.389/0	0.444/0	1.831/2
σ	1.632	1.672	1.425	1.103	1.770	0.927	1.080
Importance							
Do Not Use	873	468	632	1,119	360	1,114	259
Not Important	128	87	125	114	68	148	105
Slightly Important	198	277	323	132	175	157	314
Moderately Important	158	291	228	93	227	79	376
Important	122	264	162	58	381	32	344
Very Important	73	170	75	32	341	19	143
Mean/Mode	1.193/0	2.197/0	1.604/0	0.768/0	2.789/4	0.595/0	2.565/3
σ	1.588	1.752	1.602	1.270	1.859	1.115	1.548

Self-Generated by Author using Microsoft Word.

A review of the age variable in the bivariate regression results shows that for all statistically significant Internet-enabled communication technologies, the importance of the technology to communicate with other legislators decreases as a function of legislator age. This finding is consistent with the frequency of use and importance regressions for age shown in tables 5.3 through 5.5 which highlight that the use and importance of Internet-enabled communication technologies decreases with increasing legislator age. Another interesting general relationship highlighted in these bivariate regressions is that every statistically significant finding associated with professional legislatures suggests that professional legislators use all Internet-enabled communication technologies more frequently than do their citizen legislator counterparts. This increase in frequency of use is not surprising since, in general, professional legislators are full-time legislators, while citizen legislators are part-time legislators. Gerrymandering measures produced mixed results, with some measures indicating increased frequency of use of Internet-enabled communication technologies to communicate with constituents and other measures indicating decreased frequency of use of Internet-enabled communication technologies to communicate with their constituents. In a final general observation of the bivariate regression results, in every statistically significant result, an increase in legislator Delegate behavior is correlated with an increase in the frequency of constituent communications via Internet-enabled communication technologies. This is hardly surprising since one might expect a correlation between the frequency of communications with constituents and legislators acting as Delegates. With these general observations completed, let's examine the full regression models for each Internet-enabled communication technology.

Text Message Communications

The frequency of use of text messages with constituents is statistically significant with respect to legislator age, years in office, Senate membership, polarization, all gerrymandering measures except Convex Hull, professional legislature status (two measures), and the Trustee/Delegate scale. The frequency of text message communications with constituents was negatively correlated with legislator age. All other statistically significant variables were positively correlated. Compared to Representatives, Senators were 76 percent more likely to choose "Use Hourly" to communicate with constituents via text messages over all other frequency of use categories for text messages, on average, all else equal. The Reock gerrymandering variable indicates that a one unit increase in the Reock gerrymandering score (decreased gerrymandering) is associated with a 5.3 percent increase in the probability of a legislator indicating "Use Hourly" to communicate with constituents via text messages over all other frequency of use categories for text messages, on average, all else equal.

E-mail Communications

Statistically significant bivariate regressions associated with the frequency of use of e-mail with constituents include legislator age, years in office, political ideology (progressive to conservative), polarization, gender, education, race (black compared to white), the Convex Hull and Reock gerrymandering variables, and both professional legislature measures. Negative correlations were found with age, years in office, political ideology, and gender. Positive correlations were found with polarization, education, race, the Reock gerrymandering measure, and both professional legislature measures. A one-year increase in education is associated with a 5.8 percent increase in the probability that a legislator will choose "Use Hourly" to communicate with constituents via e-mail communication technology over all other frequency of use categories for e-mail communications, on average, all else equal.

Twitter™ Communications

Twitter™ communications with constituents were statistically significantly related to legislator age, years in office, political party ideology, both polarization binaries, gender, the Convex Hull gerrymandering measure, professional legislature status, and the Trustee/Delegate scale. Negative correlations occurred with legislator age, years in office, political ideology, gender, and the Convex Hull gerrymandering measure. Positive correlations are associated with both polarization measures, professional legislature status, and the Trustee/Delegate scale. With respect to legislator age, a one-year increase in legislator age is associated with a 5.9 percent decrease in the probability that legislators will choose "Use Hourly" to communicate with constituents via Twitter™ over all other frequency of use categories for Twitter™ communications, on average, all else equal. With respect to legislator years in office, a one-year increase in legislator years in office is associated with a 3.5 percent decrease in the probability that legislators will choose "Use Hourly" to communicate with constituents via Twitter™ over all other frequency of use categories for Twitter™ communications, on average, all else equal.

Facebook™ Communications

With respect to the frequency of use of Facebook™ communications with constituents, the statistically significant bivariate relationships include legislator age, years in office, both polarization measures, gender, and all of the gerrymandering measures except the Convex Hull measure. Negative correlations were found between the frequency of Facebook™ use with constituents and legislator age, years in office, and gender. Positive correlations were found between the frequency of Facebook™ use with constituents and both polarization variables, and all gerrymandering measures except the Convex Hull measure. With respect to the Schwartzberg gerrymandering measure, a one unit increase in the Schwartzberg gerrymandering measure (a decrease in gerrymandering) is associated with a 1 percent increase in the probability that a legislator will choose "Use Hourly" to communicate with constituents via Facebook™ over all other frequency of use categories for Facebook™ communications, on average, all else equal. With respect to gender, compared to female legislators, male legislators are 19.6 percent less likely to choose "Use Hourly" to communicate with constituents via Facebook™ over all other frequency of use categories for Facebook™ communications, on average, all else equal.

YouTube™ Communications

The use of YouTube™ to communicate with constituents is statistically significantly correlated with age, political ideology, education, and the conflict scale. Negative correlations occur with legislator age and political ideology, while positive correlation occurs with education. With respect to education, a one-year increase in legislator education is associated with a 4.2 percent increase in the probability a legislator will choose "Use Hourly" to communicate with constituents via YouTube™ over all other frequency of use categories for YouTube™ communications, on average, all else equal. With respect to legislator age, a one-year increase in legislator age is associated with a 2.4 percent decrease in the probability a legislator will choose "Use Hourly" to communicate with constituents via YouTube™ over all other frequency of use categories for YouTube™ communications, on average, all else equal.

Web Page Communications

Legislator use of web pages to communicate with constituents is statistically significantly associated with legislator age, Senate membership, political ideology, legislator education, all gerrymandering measures, both professional legislature measures, and the Trustee/Delegate scale. Positively correlated measures include Senate membership, political ideology, all gerrymandering measures, both polarization measures, and the Trustee/Delegate scale. The only negatively correlated variable is legislator age. A one-year increase in legislator age is associated with a 0.8 percent decrease in the probability that legislators will choose "Use Hourly" to communicate with constituents via web pages over all other frequency of use categories for web page communications, on average, all else equal. Compared to Representatives, Senators are 39.2 percent more likely to choose "Use Hourly" to communicate with constituents via web pages over all other frequency of use categories for web page communications, on average, all else equal.

Blog Communications

With respect to the frequency of use of blog communications to communicate with constituents, statistically significant correlations occurred with legislator age, political ideology, polarization, race (black legislators compared to white legislators), and professional legislature status. Positive correlations occurred with polarization, race, and professional legislature status while negative correlations occurred with political party. With respect to political ideology, a one unit increase in political conservativism is associated with a 4.7 percent decrease in the probability that a legislator will choose "Use Hourly" to communicate with constituents via blogs over all other frequency of use categories for blogs, on average, all else equal. Compared to white legislators, black legislators are 171.5 percent more likely to choose "Use Hourly" to communicate with constituents via blogs over all other frequency of use categories for blogs, on average, all else equal.

Constituent Communication Technology Frequency of Use Models

Three models were created to determine the relationships between constituent frequency of use of Internet-enabled communication technology and DPI variables. These models use a dependent variable *cfreqiect* which is the average frequency of use of all seven Internet-enabled communication technologies when used to communicate with constituents. In effect, these models look at the overall frequency of use for Internet-enabled communication technologies. The basic model contains legislator demographic information, the intermediate model adds primary political and institutional variables, and the advanced model adds secondary institutional variables. Additionally, the results of these three models are not presented in a table, but rather, are discussed in the paragraphs following the models presented below:

Basic Model: $cfreqiect = \beta_0 + \beta_1 (age) + \beta_2 (race) + \beta_3 (male) + \beta_4 (education) + \mathcal{E}$.

The basic model is statistically significant for age (odds ratio = 0.951, $z \leq 0.001$), and gender (odds ratio = 0.695, $z \leq 0.001$). Interpreting a single variable in the basic model for clarity, a one-year increase in legislator age is associated with a 4.9 percent decrease in the likelihood a legislator will select "Use Hourly" for the use of Internet-enabled communications with constituents over all other communication technology frequency of use categories, ceteris paribus.

Intermediate Model: $cfreqiect = \beta_0 + \beta_1(age) + \beta_2(race) + \beta_3(male) + \beta_4(education) + \beta_5(conservativism) + \beta_6(minority\ status) + \beta_7(years\ in\ office) + \beta_8(senate) + \mathcal{E}$.

The intermediate model is statistically significant for age (odds ratio = 0.951, $z \leq 0.001$) and gender (male) (odds ratio = 0.718, $z \leq 0.001$). Interpreting a single variable in the intermediate model for clarity, compared to female legislators, male legislators are 28.2 percent less likely to select "Use Hourly" for the use of Internet-enabled communications with constituents over all other communication technology frequency of use categories, ceteris paribus.

Full Model: $cfreqiect = \beta_0 + \beta_1(age) + \beta_2(race) + \beta_3(male) + \beta_4(education) + \beta_5(conservativism) + \beta_6(minority\ status) + \beta_7(years\ in\ office) + \beta_8(senate) + \beta_9(gerrymandering) + \beta_{10}(professional\ legislature\ binary) + \beta_{11}(term\ limits) + \beta_{12}(conflict\ scale) + \beta_{13}(Trustee/Delegate\ scale) + \beta_{14}(polarization\ binary) + \mathcal{E}$.

The full model is statistically significant for age (odds ratio = 0.949, $z \leq 0.001$), gender (male) (odds ratio = 0.754, $z \leq 0.05$), political party ideology (odds ratio = 0.960, $z \leq 0.05$), Senate membership (odds ratio = 1.437, $z \leq 0.01$), gerrymandering (odds ratio = 1.004, $z \leq 0.05$), professional legislature (odds ratio = 1.449, $z \leq 0.01$), and a legislator's perception of policy conflict in their district (odds ratio = 1.008, $z \leq 0.05$). Interpreting a single variable in the full model for clarity, a legislator being a member of a professional legislature is associated with a 44.9 percent increase in the likelihood a legislator will select "Use Hourly" for the use of Internet-enabled communications with constituents over all other communication technology frequency of use categories, ceteris paribus.

The previous section examines frequency of use models using an average of all seven Internet-enabled communication technologies. Table 4.5 contains the results of the full regression model using *each* individual Internet-enabled communication technology as the dependent variable.

A review of the age variable in table 4.5 shows that for all statistically significant age variables associated with each Internet-enabled communication technology model, the frequency of use of the technology to communicate with constituents decreases as a function of legislator age. This finding is

Table 4.5 Constituent Frequency of Use DPI Full Regression Models, Odds Ratios

Political Party (progressive to conservative)	0.981	0.882***	0.949*	0.983	0.923**	1.091***	0.956
Polarization Binary	1.555**	1.647**	1.426*	0.930	0.968	0.894	1.277
Male	0.959	0.696*	0.654**	0.809	1.079	1.090	1.193
Education	0.997	1.058*	1.013	1.040	1.063*	0.975	0.986
Race (Compared to White)	Omitted Category: White	Omitted Category: White	Omitted Category: White	Omitted Category: White	Omitted Category: White	Omitted Category: White	Omitted Category: White
Black	1.090	1.242	0.955	1.598	1.102	1.836*	2.445**
Other	0.886	0.677	0.609**	0.990	0.675	1.135	1.105
Gerrymandering (Comprehensive)	1.006***	0.999	0.997	1.005**	0.999	1.006***	1.000
Minority One Chamber	0.946	0.901	1.191	1.210	0.663	1.024	0.653
Minority Two Chambers	0.945	0.738	0.975	1.202	0.997	0.997	0.619
Minority Two Chambers and Governorship	1.121	0.842	0.936	1.103	0.583**	1.037	0.959
Professional Legislature Binary	1.555**	1.871***	1.055	0.930	0.951	1.579**	1.220
Term Limits	0.790	0.623***	0.950	0.876	0.952	1.031	0.770
Conflict Scale	1.000	1.011**	1.009*	1.004	1.011*	1.006	1.002
Trustee/Delegate Scale	1.005***	1.000	1.002	1.000	1.003	1.004*	1.000

* z ≤.05, ** z ≤.01, *** z ≤.001.
Self-Generated by Author using Microsoft Word.

consistent with the frequency of use and importance bivariate and multivariate regression modeling of Internet-enabled communication technologies examined in previous regression models. Another interesting general relationship highlighted in table 4.5 is that every statistically significant finding associated with legislator gender shows that male legislators use Internet-enabled communication technologies less frequently when communicating with constituents than female legislators. For all statistically significant results shown in table 4.5, legislator polarization was correlated with increases in the frequency of use of Internet-enabled communication technologies to communicate with constituents. A legislator's perception of policy conflict between constituents in a legislator's district is always associated with an increase in the use of the Internet-enabled communication technology to communicate with their constituents. One final general observation is that once again, professional legislators communicate more frequently with their constituents via Internet-enabled communication technologies than do their citizen legislator counterparts. With these general observations complete, let's examine the individual Internet-enabled communication technology frequency of use with constituent's full regression models.

Text Message Communications

The frequency of use of text messages to communicate with constituents is statistically significantly associated with legislator age, years in office, Senate membership, the binary polarization measure, the comprehensive gerrymandering measure, the binary professional legislature measure, and the Trustee/Delegate scale. Positive correlations occur with years in office, Senate membership, the polarization binary, the comprehensive gerrymandering measure, the professional legislature binary variable, and the Trustee/Delegate scale. With respect to years in office, a one-year increase in the length of time a legislator is in office is associated with a 3.3 percent increase in the probability that a legislator will choose "Use Hourly" for text message communications with constituents over all other text message frequency of use categories, on average, all else equal. With respect to polarization, legislators who self-identify as strongly progressive or strongly conservative are 55.5 percent more likely as their more moderate peers to choose "Use Hourly" for text message communications with constituents over all other text message frequency of use categories, on average, all else equal.

E-mail Communications

The frequency of use of e-mail communications with constituents is statistically significantly associated with legislator age, Senate membership, political ideology, the binary polarization measure, the binary professional legislature measure, gender, education, term limits, and the conflict scale. Positive correlations occur with Senate membership, education, the professional legislator binary measure, and the conflict scale. Legislator age, political ideology, gender, and term limits are negatively correlated. With respect to polarization, legislators who self-identify as strongly progressive or strongly conservative are 64.7 percent more likely as their more moderate peers to choose "Use Hourly" for e-mail communications with constituents over all other e-mail frequency of use categories, on average, all else equal.

Twitter™ Communications

The frequency of use of Twitter™ communications with constituents is statistically significantly associated with legislator age, years in office, Senate membership, political ideology, the binary polarization measure, the binary professional legislature measure, gender, the other race category compared to whites, and the conflict scale. The only positively correlated variables are Senate membership and the conflict scale; all other variables are negatively correlated. With respect to years in office, a one-year increase in the length of time a legislator is in office is associated with a 3.3 percent decrease in the probability that a legislator will choose "Use Hourly" for Twitter™ communications with constituents over all other Twitter™ frequency of use categories, on average, all else equal. With respect to polarization, legislators who self-identify as strongly progressive or strongly conservative are 43.8 percent more likely as their more moderate peers to choose "Use Hourly" for Twitter™ communications with constituents over all other Twitter™ frequency of use categories, on average, all else equal.

Facebook™ Communications

The frequency of use of Facebook™ communications with constituents is statistically significantly associated with legislator age (negatively correlated) and the comprehensive gerrymandering measure (positively correlated). With respect to age, a one-year increase in legislator age is associated with a 2.3 percent increase in the probability that a legislator will choose "Use Hourly" for Facebook™ communications with constituents over all other Facebook™ frequency of use categories, on average, all else equal. A one unit increase in gerrymandering score (less gerrymandering) was associated with a 0.5 percent increase in the likelihood that a legislator would choose "Use Hourly" for Facebook™ communications with constituents over all other Facebook™ frequency of use categories, on average, all else equal.

YouTube™ Communications

The frequency of use of YouTube™ communications with constituents is statistically significantly associated with legislator age, political ideology, education, minority party status in two legislative chambers and the governorship, and the conflict scale. Education and the conflict scale are positively correlated, while all other statistically significant measures are negatively correlated. With respect to age, a one-year increase in legislator age is associated with a 2.6 percent increase in the probability that a legislator will choose "Use Hourly" for YouTube™ communications with constituents over all other YouTube™ frequency of use categories, on average, all else equal. With respect to education, a one-year increase in education is associated with a 6.3 percent increase in the probability that a legislator will choose "Use Hourly" for YouTube™ communications with constituents over all other YouTube™ frequency of use categories, on average, all else equal.

Web Page Communications

The frequency of use of web page communications with constituents is statistically significantly associated with legislator age, political ideology, race (black legislators compared to white legislators), the comprehensive gerrymandering measure, the professional legislature binary variable, and the Trustee/Delegate measure. Age is negatively correlated and all other statistically significant measures are positively correlated. With respect to age, a one-year increase in legislator age is associated with a 1 percent increase in the probability that a legislator will choose "Use Hourly" for web page communications with constituents over all other web page frequency of use categories, on average, all else equal. A one unit increase in gerrymandering score (less gerrymandering) was associated with a 0.5 percent increase in the likelihood that a legislator would choose "Use Hourly" for web page communications with constituents over all other web page frequency of use categories, on average, all else equal.

Blog Communications

The frequency of use of blog communications with constituents is statistically significantly associated with legislator age (negatively correlated) and race (black legislators compared to white legislators, positively correlated). With respect to age, a one-year increase in legislator age is associated with a 3.61 percent increase in the probability that a legislator will choose "Use Hourly" for blog communications with constituents over all other blog frequency of use categories, on average, all else equal. Compared to white legislators, black legislators are 145.5 percent more likely to choose "Use Hourly" for blog communications with constituents over all other blog frequency of use categories, on average, all else equal.

Internet-Enabled Communication Technology Importance with Constituents

Bivariate ordinal logistic regressions were run for the importance of Internet-enabled communications when communicating with constituents, and DPI variables. The results of these importance ordinal bivariate regressions are summarized in the sections below.

A review of the age variable in the bivariate regressions shows that for all statistically significant age variables associated with each Internet-enabled communication technology model, the importance of Internet-enabled communication technology to communicate with constituents decreases as a function of legislator age. This finding is consistent with the other frequency of use and importance bivariate and multivariate regression modeling of Internet-enabled communication technologies examined in this chapter. Another interesting general relationship highlighted in these bivariate regressions is that every statistically significant finding associated with legislator gender shows that male legislators find Internet-enabled communication technologies less important when communicating with constituents than do female legislators.

For all statistically significant results, legislator polarization was correlated with increases in the importance of Internet-enabled communication technologies to communicate with constituents. Compared to white legislators, black legislators find Internet-enabled communication technologies more important when communicating with their constituents. One final general observation is that once again, professional legislators find Internet-enabled communication technologies more important to communicate with their constituents than do their citizen legislator counterparts. With these general observations complete, let's examine the individual Internet-enabled communication technology importance bivariate regression models.

Text Message Communications

Statistically significant relationships correlated with the importance of text message communications with constituents include legislator age, years in office, Senate membership, polarization, the Reock gerrymandering measure, and professional legislature status. All measures other than age have positive correlations. With respect to age, a one-year increase in legislator age is associated with a 1.6 percent decrease in the probability that a legislator will select "Very Important" for text message communications with constituents over all other text message importance categories, on average, all else equal. Compared to Representatives, Senators are 41.1 percent more likely to select "Very Important" for text message communications with constituents over all other text message importance categories, on average, all else equal

E-mail Communications

Statistically significant relationships correlated with the importance of e-mail communications with constituents include legislator age, years in office, political party ideology, polarization, gender, education, race (black legislators compared to white legislators), gerrymandering (two measures), and professional legislature status (two measures). Positive correlations include both polarization measures, education, race (black legislators compared to white legislators), and both professional legislature measures. Negative correlations include legislator age, years in office, political party ideology, and gender. With respect to age, a one-year increase in legislator age is associated with a 5.9 percent decrease in the probability that a legislator will select "Very Important" for e-mail communications with constituents over all other e-mail importance categories, on average, all else equal. A one unit increase in conservative political party ideology is associated with a 9.3 percent decrease in the probability that a legislator will select "Very Important" for e-mail communications with constituents over all other e-mail importance categories, on average, all else equal.

Twitter™ Communications

Statistically significant relationships correlated with the importance of Twitter™ communications with constituents include legislator age, years in office, political party ideology, polarization, gender, race (black legislators compared to white legislators), the Convex Hull gerrymandering measure, and professional legislature status. Positive correlations include polarization, race, and the professional legislature measure. Negative correlations include age, years in office, political party ideology, gender, and the Convex Hull gerrymandering measure. With respect to age, a one-year increase in legislator age is associated with a 5 percent decrease in the probability that a legislator will select "Very Important" for Twitter™ communications with constituents over all other Twitter™ importance categories, on average, all else equal. A one unit increase in conservative political party ideology is associated with a 4.4 percent decrease in the probability that a legislator will select "Very Important" for e-mail communications with constituents over all other e-mail importance categories, on average, all else equal.

Facebook™ Communications

Statistically significant relationships correlated with the importance of Facebook™ communications with constituents include legislator age, years in office, political party ideology, polarization, gender, race (black legislators compared to white legislators), the Reock and Polsby-Popper gerrymandering measures, and professional legislature status. Positive correlations include both polarization measures, race (black legislators compared to white legislators), the Reock gerrymandering measure, and the professional legislature measure. With respect to age, a one-year increase in legislator age is associated with a 2.7 percent decrease in the probability that a legislator will select "Very Important" for Facebook™ communications with constituents over all other Facebook™ importance categories, on average, all else equal. A one unit increase in conservative political party ideology is associated with a 2 percent decrease in the probability that a legislator will select "Very Important" for e-mail communications with constituents over all other e-mail importance categories, on average, all else equal.

YouTube™ Communications

Statistically significant relationships correlated with the importance of YouTube™ communications with constituents include legislator age, years in office, political party ideology, gender, and race (black legislators compared to white legislators). All correlations are negative except race. With respect to age, a one-year increase in legislator age is associated with a 3.5 percent decrease in the probability that a legislator will select "Very Important" for YouTube™ communications with constituents over all other YouTube™ importance categories, on average, all else equal. A one unit increase in conservative political party ideology is associated with a 3.5 percent decrease in the probability that a legislator will select "Very Important" for e-mail communications with constituents over all other e-mail importance categories, on average, all else equal.

Web Page Communications

Statistically significant relationships correlated with the importance of web page communications with constituents include legislator age, political party ideology, education, all gerrymandering measures except Convex Hull, and professional legislature status (both measures). All correlations are positive except legislator age and education. With respect to age, a one-year increase in legislator age is associated with a 1.3 percent decrease in the probability that a legislator will select "Very Important" for web page communications with constituents over all other web page importance categories, on average, all else equal. A one unit increase in conservative political party ideology is associated with a 1.3 percent decrease in the probability that a legislator will select "Very Important" for e-mail communications with constituents over all other e-mail importance categories, on average, all else equal.

Blog Communications

Statistically significant relationships correlated with the importance of blog communications with constituents include legislator age, political party, gender, and race (black legislators compared to white legislators). All correlations are negative except for race. With respect to age, a one-year increase in legislator age is associated with a 3.7 percent decrease in the probability that a legislator will select "Very Important" for blog communications with constituents over all other blog importance categories, on average, all else equal. A one unit increase in conservative political party ideology is associated with a 3.7 percent decrease in the probability that a legislator will select "Very Important" for e-mail communications with constituents over all other e-mail importance categories, on average, all else equal.

Constituent Communication Technology Importance Models

Three models were created to determine the relationships between the importance of Internet-enabled communication technology and DPI variables. These models use a dependent variable *cimportiect* which is the average of all seven Internet-enabled communication technologies when used to communicate with constituents. In effect, these models look at the overall importance of Internet-enabled communication technology. The basic model contains legislator demographic information, the intermediate model adds primary political and institutional variables, and the advanced model adds secondary institutional variables. Additionally, the results of these three models are not presented in a table, but rather, are discussed in the paragraphs following the models presented below:

Basic Model: $cimportiect = \beta_0 + \beta_1(age) + \beta_2(race) + \beta_3(male) + \beta_4(education) + \mathcal{E}.$

The basic model is statistically significant for age (odds ratio = 0.951, $z \leq 0.001$), black legislators compared to white legislators (odds ratio = 2.110, $z \leq 0.01$), and gender (odds ratio = 0.628, $z \leq 0.001$). Interpreting a single variable in the basic model for clarity, a one-year increase in legislator age is associated with a 1 percent increase in the likelihood a legislator will select "Very Important" for the use of Internet-enabled communications with constituents over all other communication technology importance categories, ceteris paribus.

Intermediate Model: *cimportiect* $= \beta_0 + \beta_1(age) + \beta_2(race) + \beta_3(male) + \beta_4(education) + \beta_5(conservativism) + \beta_6(minority\ status) + \beta_7(years\ in\ office) + \beta_8(senate) + \mathcal{E}$.

The intermediate model is statistically significant for age (odds ratio = 0.952, $z \leq 0.001$), black legislators compared to white legislators (odds ratio = 1.910, $z \leq 0.05$), gender (odds ratio = 0.635, $z \leq 0.0001$), and Senate (odds ratio = 1.390, $z \leq 0.01$). Interpreting a single variable in the intermediate model for clarity, compared to Representatives, Senators are 39 percent more likely to select "Very Important" for the use of Internet communications with constituents over all other communication technology importance categories, ceteris paribus.

Full Model: *cimportiect* $= \beta_0 + \beta_1(age) + \beta_2(race) + \beta_3(male) + \beta_4(education) + \beta_5(conservativism) + \beta_6(minority\ status) + \beta_7(years\ in\ office) + \beta_8(senate) + \beta_9(gerrymandering) + \beta_{10}(professional\ legislature\ binary) + \beta_{11}(term\ limits) + \beta_{12}(conflict\ scale) + \beta_{13}(Trustee/Delegate\ scale) + \beta_{14}(polarization\ binary) + \mathcal{E}$.

The full model is statistically significant for age (odds ratio = 0.951, $z \leq 0.001$), gender (odds ratio = 0.691, $z \leq 0.01$), Senate (odds ratio = 1.370, $z \leq 0.05$), gerrymandering (odds ratio = 1.004, $z \leq 0.05$), Delegate behavior (odds ratio = 1.003, $z \leq 0.05$), and the polarization measure (odds ratio = 1.337, $z \leq 0.01$). Interpreting a single variable in the full model for clarity, a 1 percent increase in a legislator's behavior as a Delegate is associated with a 0.3 percent increase in the likelihood a legislator will select "Very Important" for the use of Internet-enabled communications with constituents over all other communication technology importance categories, ceteris paribus.

The previous section examines communication technology importance models using an average of all seven Internet-enabled communication technologies. Table 4.6 contains the results of the full regression model using *each* individual Internet-enabled communication technology as the dependent variable.

Text Message Communications

In the full regression model for text messages, age, years in office, gerrymandering, term limits, and the Delegate scale are all statistically significant. Age and term limits are associated with a decrease in the importance of text message use to communicate with constituents, while years in office, Senate membership, gerrymandering, and the Trustee/Delegate scale are associated with an increase in the importance of text messages. The largest effect size is associated with Senate membership, where Senate membership is associated

Table 4.6 Constituent Importance DPI Full Regression Models Odds Ratios TBD

Constituent Importance	*Text Messages*	*e-mail*	*Twitter™*	*Facebook™*	*YouTube™*	*Web Pages*	*Blogs*
Age	0.980***	0.940***	0.955***	0.974***	0.963***	0.988*	0.957***
Years in Office	1.020*	0.987	0.969***	0.992	0.991	1.003	1.003
Senate (Compared to House)	1.588***	1.308	1.211	1.132	1.007	1.270	1.209
Political Party (progressive to conservative)	0.999	0.886***	0.952*	0.977	0.965	1.083***	1.000
Polarization Binary	1.146	1.815***	1.350	1.260	1.272	1.465*	1.188
Male	1.017	0.659**	0.647***	0.799	0.695*	0.924	0.890
Education	0.979	1.039	1.000	1.024	1.036	0.979	1.010
Race (Compared to White)	Omitted Category: White	Omitted Category: White	Omitted Category: White	Omitted Category: White	Omitted Category: White	Omitted Category: White	Omitted Category: White
Black	1.283	1.284	1.167	1.270	1.679	1.062	3.358***
Other	0.941	0.772	0.693*	1.025	0.896	1.066	1.198
Gerrymandering (Comprehensive)	1.004*	1.001	0.997	1.004	1.001	1.006	
Minority One Chamber	0.845	0.730	1.174	0.991	0.668	1.059	0.613
Minority Two Chambers	1.380	0.799	0.930	1.079	1.076	0.884	1.148
Minority Two Chambers and Governorship	1.239	0.848	1.018	0.935	0.866	0.965	1.127
Professional Legislature Binary	1.189	1.324	1.010	0.921	0.740	1.465*	0.952
Term Limits	0.687**	0.769	0.961	0.879	0.985	1.135	0.777
Conflict Scale	0.996	1.002	1.008*	0.997	1.008	1.000	1.001
Trustee/Delegate Scale	1.005**	1.000	1.001	1.001	1.001	1.003*	1.001

* z ≤.05, ** z ≤.01, *** z ≤.001.
Self-Generated by author using Microsoft Word.

with 58.8 percent increase in the likelihood a legislator will choose "Very Important" for text message communication over all other text message importance categories, on average, all else equal. A one unit increase in state gerrymandering (less gerrymandering) is associated with a 0.4 percent increase in the likelihood that a legislator will choose "Very Important" for text message use with constituents over all other text message importance categories, ceteris paribus.

E-mail Communications

With respect to the importance of e-mail communications with constituents, both the polarization binary variable (legislators indicating strongly conservative or strongly progressive political party affiliation were coded as 1, all other political party affiliations were coded as 0) and the ordinal polarization variable are statistically significantly related to the importance of using e-mail to communicate with constituents. Age and gender are associated with a decrease in the importance of e-mail use to communicate with constituents. A one-year increase in age is associated with a 6 percent decrease in the probability that a legislator will indicate that e-mails are very important to communicate with their constituents. Compared with female legislators, male legislators are 34.1 percent less likely to choose "Very Important" for e-mail use with constituents over all other e-mail importance categories, on average, all else equal.

Twitter™ Communications

The Twitter™ model has statistically significant negative relationships with age, years in office, ordinal polarization variable, gender, and race, and a positive relationship with the conflict scale measure. Compared to female legislators, male legislators are 35.3 percent less likely to choose "Very Important" over all other Twitter™ importance categories when using Twitter™ to communicate with constituents, on average, all else equal. Compared with white legislators, other races are 30.7 percent less likely to choose "Very Important" over all other Twitter™ importance categories when using Twitter™ to communicate with constituents, ceteris paribus.

Facebook™ Communications

Only legislator age has a statistically significant relationship with Facebook™ use with constituents. A one-year increase in age is associated with a 2.6 percent decrease in the probability that a legislator will indicate that Facebook™ is very important to communicate with their constituents, on average, all else equal.

YouTube™ Communication

The YouTube™ model has statistically significant relationships with age and gender. A one-year increase in age is associated with a 3.7 percent decrease in the probability that a legislator will indicate that YouTube™ is very important to communicate with their constituents over all other YouTube™ importance categories, on average, all else equal. Compared to female legislators, male legislators are 30.5 percent less likely to choose "Very Important" for using Twitter™ to communicate with constitutes over all other YouTube™ importance categories, on average, all else equal.

Web Page Communication

The importance of web pages to communicate with constituents is associated with legislator age, political ideology (progressive to conservative scale), polarization, professional legislature status, and Trustee/Delegate scale. Only age is negatively correlated with the importance of using web pages to communicate with constituents, while all other statistically significant variables are positively correlated. A one-year increase in legislator age is associated with a 1.2 percent decrease in the probability that a legislator will choose "Very Important" for web page communications with constituents over all other web page communication importance categories, on average, all else equal. Compared to citizen legislators, professional legislators are 46.5 percent more likely to choose "Very Important" for web page communications with constituents over all other web page communication importance categories, on average, all else equal.

Blogs Communication

The importance of blog communications with constituents has two statistically significant relationships: age and race. A one-year increase in age is associated with a 4.3 percent decrease in the probability that a legislator will indicate that blogs are very important to communicate with their constituents over all other blog importance categories, on average, all else equal. Compared to white legislators, black legislators are 235.8 percent more likely to choose "Very Important" for using blogs to communicate with their constituents over all other blog importance categories, on average, all else equal.

CONCLUSION

The qualitative research in this chapter uncovered legislator behaviors that were both expected and unexpected. Among the expected findings are the following: (1) legislators varied significantly in their use of communication technology; (2) legislators confirm the quantitative research on mature communication technology by verifying in their own words that more natural CTs are more important to them, primarily because natural CTs tend to convey information such as emotions and honesty/deception better than less natural CTs; and (3) demographic relationships between communication technology frequency of use and importance obtained during the quantitative research were confirmed. Significantly, communication technology frequency of use and importance rankings identified in the quantitative research were significantly correlated with the frequency of discussion for each communication technology discussed during qualitative interviews. Network diagrams discussed in this chapter highlight the complexities surrounding communication technology, complexities that cannot be exposed by quantitative research. For example, qualitative research shows that legislators prefer not to use e-mail with constituents out of concern for FOIA laws, while quantitative research shows that e-mail frequency of use with constituents was ranked tenth out of fifteen and e-mail frequency of use with peers was ranked first. Put succinctly, quantitative research exposes the relationship, while qualitative research helps explain why the relationship exists. This is but one of many possible examples.

More interesting than the expected findings were the unanticipated findings uncovered during qualitative research. These findings include the following: (1) there are significant mismatches between the communication technologies that legislators find most important and the communication technologies that constituents use most frequently; (2) legislators are using e-mail to avoid communications from constituents who disagree with their policy and/or political ideology and using e-mail to enhance communications with constituents who agree with their policy and/or political ideology; (3) legislators are using text messaging real-time during floor debates to obtain outside information to assist them as they debate other legislators; (4) communication technology risk perceptions are related to the naturalness of the communication technology, with more natural communication technologies involving less perceived risk for the legislator; and (5) legislators use communication technologies as a challenge avoidance mechanism; legislators are more likely to engage with constituents with opposing ideologies if the engagement is face to face, slightly less likely to engage these constituents over the phone, and unlikely to engage these constituents via e-mail. Interviews suggest that naturalness theory offers an explanation for this behavior. Legislators believe they are unlikely to convince a constituent to change their mind about a topic via e-mail, slightly more likely to do so over the phone, and two legislators indicated they have changed the minds of every constituent they have ever met face-to-face.

Younger legislators are driving CT infrastructure development and changes, while older legislators are resisting these changes. In addition, technology itself is driving reductions in IT support staffing levels through automation made possible by new technologies. In an example of younger legislators driving changes in IT support infrastructure, several younger legislators requested IT support in setting up personal cell phones and computers to access the Arizona capitol intranet. These requests required secure tunneling VPN technology that was not in use by the IT department. The request was sent to the director and approved, and now, setting up legislators' private computers and cell phones is one of the IT departments' most requested services. In another example, legislators requested bills be saved in PDF format in addition to Microsoft Word format.

Interviewee IT A categorized legislators into three fundamental categories: legislators who do not want to use technology at all, and have their legislator assistants deal with technology, legislators "in the middle" who use the technology but do not demand or embrace it, and finally, the third group of legislators who demand new communication technology. IT A indicated that older legislators tend to fall in the first category, while younger legislators tend to fall in the third category. IT B calls the first category of legislators "cowboy legislators" while noting that these older legislators wear cowboy boots under their suits.

Legislators are driving IT infrastructure changes and IT staff are adjusting their roles to support legislator requests. On the other hand, technology itself is driving reductions to IT staff by automating tasks that were once assigned to an IT staffer. IT A provides the example of web page creation. Web page creation was once a manual process and now it is automated through the use of scripts and custom programs. These factors suggest that IT infrastructure is a dynamic environment where technology can both add to the task load of IT personnel as well as reduce it. Interviews with IT staff suggest that legislators do in fact precipitate changes in IT infrastructure and support.

From a quantitative perspective, there are several important themes which occurred in the analyses completed in chapter 5. First, with respect to legislator age, legislators use and value Internet-enabled communication technologies less the older they get. This phenomenon does not occur with the frequency of use of mature communication technologies. Figures 4.1 and 4.2 highlight these relationships. Figure 4.1 is a scatter plot of Internet-Enabled communication technology frequency of use as a function of legislator age, while figure 4.2 is a scatter plot of the frequency of use of mature communication technologies as a function of legislator age. Both figure 4.1 and figure 4.2 contain the best fit straight line function with 95 percent confidence interval banding.

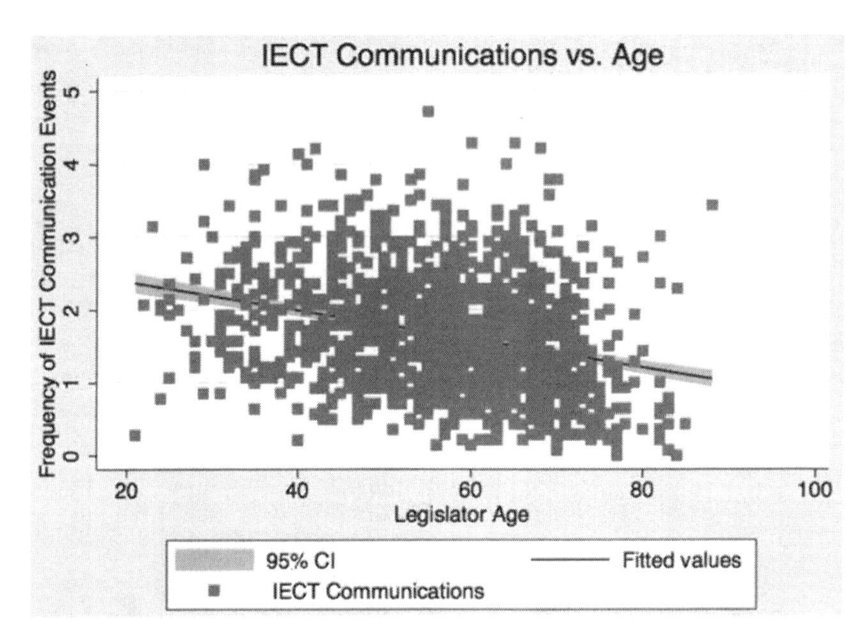

Figure 4.1 Scatter plot, Internet-enabled CT frequency of use versus age.
Self-Generated by author using Stata Quantitative Analyses Program.

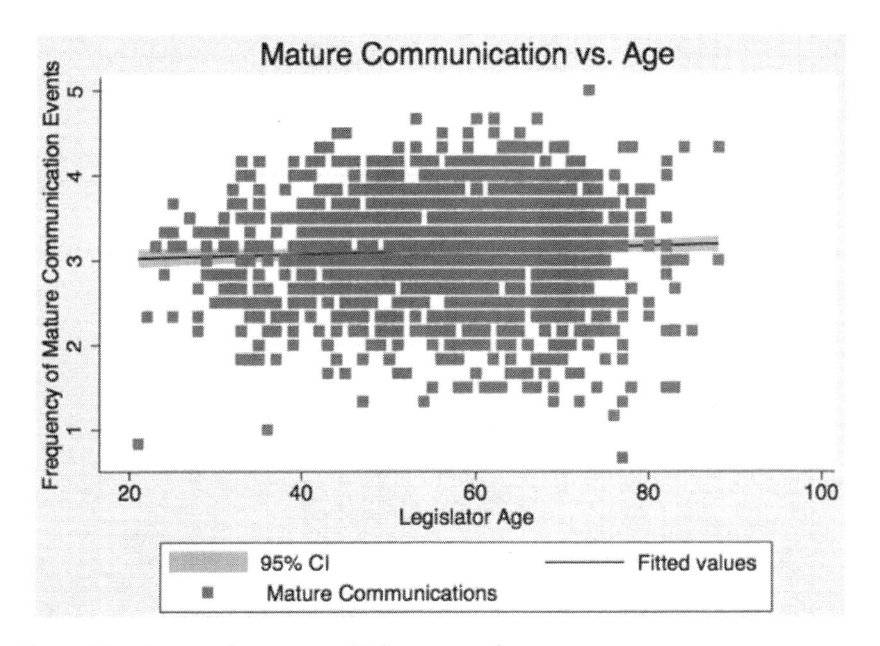

Figure 4.2 Scatter plot, mature CT frequency of use versus age.
Self-Generated by author using Stata Quantitative Analyses Program.

Second, even after controlling for the effects of age, the frequency of use and importance of Internet-enabled communication technologies when communicating with peers and constituents decreases the longer a legislator is in office. Third, male legislators use and value Internet-enabled communication technologies less than female legislators. Fourth, increases in legislator polarization are associated with both the frequency and importance of Internet-enabled communication technologies. Fifth, black legislators use and value Internet-enabled communication technologies more than their white counterparts. Sixth, increases in legislator Delegate behaviors are associated with increases in both the frequency and use of Internet-enabled communication technologies. Seventh, although the various gerrymandering measures did not provide consistent results, decreases in gerrymandering at the state level are statistically significantly correlated with increases in the frequency of use and importance of Internet-enabled communication technologies. The less safe a legislator's district is, the more likely they are to use and value Internet-enabled communication technologies. The policymaking implications of these relationships will be discussed in detail in chapter 8.

From the perspective of the critical frequency theory, Internet-enabled communication technologies offer significant evidence of the complexity surrounding both frequency of use and feedback path attenuation associated with the importance of Internet-enabled communication technologies. For example, black legislators use and value Internet-enabled communication technologies more than white legislators and male legislators use and value Internet-enabled communication technologies more than female legislators. Adding to this complexity, age decreases the frequency of use and importance of Internet-enabled communication technology to legislators. A constituent trying to communicate with an older white female legislator via an Internet-enabled communication technology may not have their communication received or treated as important when compared to the same constituent communicating with a younger black male legislator. In effect, the feedback loop is broken or significantly attenuated when a constituent communicates with a constituent via a communication feedback path that the legislator neither uses nor values.

NOTES

1. Text messaging can be direct via cellular signal without TCP/IP, but for the purposes of these analyses, both SMS and MMS formats are considered as an Internet-enabled communication technology.
2. Boilerplate e-mail is defined as an e-mail where the majority of the e-mail content is standardized by a policy advocate. For example, the National Rifle

Association's request that members fill out a predefined e-mail with their name and address, and e-mail it to their legislator. The NRA suggests the content of the e-mail and the NRA member simply provides their contact information to "legitimize" the e-mail. As we will see in this book, boilerplate e-mails are mostly ineffective tools for constituents to communicate policy preferences to legislators.

3. Along with legislator-constituent communication technology mismatches and minority party use of communication technology.

Chapter 5

Relative Use and Importance of all Communication Technologies

OVERVIEW

In each of the previous three chapters, three forms of communication technologies were examined: (1) Mature communication technologies; (2) Internet-enabled communication technologies; and (3) Mass-media communication technologies. This chapter brings all three of these forms of communication technologies together to understand their relative relationships. Examination begins with the importance of the various communication technologies. Mass-media communication technologies are compared against mature and Internet-enabled communication technologies with respect to frequency of use and importance, but cannot (like mature and Internet-enabled communication technologies) be compared to constituent use vs. peer use; no attempt was made to instrument mass-media communication technology frequency of use or importance when used to communicate with other legislators; it was assumed that legislators do not, in general, use mass-media communication technologies to communicate with other legislators, with the possible exception of policy positioning.

IMPORTANCE OF COMMUNICATION TECHNOLOGIES

Legislators interviewed by West's (2014) research into legislator use of communication technologies found that legislators broadly categorized communication technology into two functional forms: (1) communication technologies associated with campaigning and communicating with constituents and (2) communication technologies associated with the business of legislating (100). Almost all (if not all) of the communication technologies examined in this chapter can be used for both campaigning and for the business of legislating, some less than others. For example, legislators are unlikely to use blogs for the business of legislating, while face-to-face communications are useful for both the business of legislating and campaigning. Insights into each functional form can be gleaned from the data gathered for this book since the instrument examined both constituent communications and peer communications.

With respect to the relative importance of mature, Internet-enabled, and mass-media communication technologies, each form of communication technology is examined relative to all other communication technologies. This process begins with hypothesizing the relative importance of a communication technology by using theories discussed in chapter 1. As identified in table 5.1, the hypothesized theoretical importance of each communication technology is derived using three guidelines, all related to media naturalness theory: (1) The age of the technology, (2) by duplex (which is the communication real-time bidirectional [like a face-to-face conversation] or a time-based serial communication [like a written letter, e-mail, or blog post]), and finally, (3) by media bandwidth. The final sort by media bandwidth is a recognition of the value of media richness theory which proposes that richer

Table 5.1 Importance of Communication Technologies Used by Legislators

Hypothesized Importance	*Communication Technology*	*Naturalness*
1	Face-to-Face Communications	Oldest form of communication. Full-duplex, verbal and visual cues available. Very mature communication technology.
2	Phone Conversations	Speaking is the oldest form of communication but the phone is a newer (relatively) technology, full-duplex, moderate bandwidth.
3	Non-electronic Written Communications	Second oldest form of communication. Half-duplex, low bandwidth.
4	e-mail	1971. Half-duplex, low bandwidth.
5	Web Pages	1989. Half-duplex, higher bandwidth.
6	Text Messages	1992. Half-duplex, emoticons available to cue meaning.
7	Blogs	1994. Half-duplex, moderate bandwidth.
8	Facebook™	2004. Half-duplex, high bandwidth.
9	YouTube™	2005. Half-duplex, high bandwidth
10	Twitter™	2006. Half-duplex, low bandwidth.

Self-Generated by author using Microsoft Word.

media consume a larger electronic bandwidth and convey more information (Burke and Chidambaram, 1999).

E-mail is assumed to be used primarily without attachments that contain visual or audio cues, and blogs, web pages, and Twitter™ while they can be half-duplex, are primarily unidirectional in nature. Table 5.1 assumes that the importance of communicating with constituents and peers is driven by media naturalness theory and not by the mass distribution capabilities of each communication technology. A communication technology's mass distribution characteristics are not considered in determining the importance of a communication technology in table 5.1. Note that table 5.1 does not list the theoretical importance of any of the mass-media communication technologies, primarily because it is difficult to quantify the mass distribution characteristics which are market dependent. For example, in the rural southeastern community of Laurinburg North Carolina there may be a radio listening market of 10 percent of the population (2017), or 1,596 listeners. Compare this to 10 percent of New York's population, or 862,270 (2017) listeners, and the difference in the importance of radio becomes clearer.

In the previous section, the expected *relative* importance of communication technology to legislators was derived using media richness, media naturalness, and the electrical engineering concepts of bandwidth and duplex. These results can be compared to the measured relative importance rankings shown in table 5.2 While no research could be found which directly measures the relative importance legislators assign to various communication technologies, indicators of the overall importance of various communication technologies can be found. For example, Ferber et al. (2005) and West (2014) surveyed Arizona legislators and found that members "overwhelmingly prefer face-to-face communication" (Ferber et al., 2005, 149) to computer-mediated communication technologies when performing their duties as legislators. Although Ferber et al. measured overall popularity of communication technologies and not overall importance, the two concepts are likely to be linked. Ferber's findings, when compared with the hypothesized importance of communication technology in table 5.1, suggest a link between the importance of a communication technology as predicted by media naturalness theory and the popularity of communication media to legislators. Ferber et al. noted that legislators viewed face-to-face interactions as most popular (31.7 percent). Telephones were second most popular (23.1 percent), followed by e-mail (19.2 percent) and regular mail (18.4 percent). Ferber's findings, taken roughly ten years prior to the data gathered for this book, mirror the findings in this book quite closely with respect to the importance of mature communication technologies.

Table 5.2 Communication Technology Importance, Peer and Constituent Communications

Communication Technology	*Constituent Importance Ranking*	*Constituent Mean Value*	*Peer Importance Ranking*	*Peer Means and Difference of Peer and Constituent Mean t-test Results*
Face-to-Face Meetings	1	4.64	1	4.59*
Telephone	2	4.40	3	4.05***
Letters (Hardcopy)	3	4.36	5	2.64***
Automated Phone Calls	4	3.32	N/A	N/A
Press Release	5	2.91	N/A	N/A
Web Page	6	2.85	9	0.70***
Text Messages	7	2.57	10	0.54***
Twitter™	8	2.38	8	1.07***
Facebook™	9	1.63	6	1.74
Television	10	1.52	N/A	N/A
E-mail	11	1.28	2	4.31***
Town-Hall Meetings	12	0.95	N/A	N/A
Radio	13	0.78	N/A	N/A
YouTube™	14	0.60	7	1.22***
Blog	15	0.50	4	3.77***

*p< = 0.05, **p< = 0.01, ***p< = 0.001.
Self-Generated by author using Microsoft Word.

Research by Burke and Chidambaram (1999) uncovered evidence that groups *initially* found the face-to-face medium to be more effective compared to Web-based synchronous and asynchronous communications; this effectiveness differential disappeared the longer the teams communicated. This suggests that, over time, and with the experience gained from group interactions, other communication technologies may be seen to be as effective at transmitting information as face-to-face communications. This may explain the differing importance of communication technologies when legislators communicate with constituents as compared to when they are communicating with their peers. Like Ferber et al. (2005), other researchers who note a human preference for face-to-face communications over other forms of communication indicate that a preference for one communication technology over another depends on many factors. These factors include time constraints (Caballer, Gracia, and Peiró, 2005; Daft and Lengel, 1986), symbolic needs[1] (Denhardt et al., 2008; Trevino, Lengel, and Daft, 1987), and of course, the availability of the media itself for use.

With the *expected* importance of communication technologies identified in table 5.1, table 5.2 highlights the *measured* importance peer and constituent importance for the various communication technologies. The statistical significance shown in the peer mean value column reflects the difference of means between the constituent mean value for communication technology importance and the peer mean value for communication technology importance. For example, the mean value of the importance of hardcopy letters when communicating with constituents is 4.36, somewhere between important and very important, while the mean peer importance value for hardcopy letters is 2.64, falling somewhere between slightly important and moderately important. The difference of means between peer and constituent communications with hardcopy letters is statistically significant at the 99.9 percent confidence interval.

As shown in table 5.2, legislators find face-to-face meetings, telephone, and letters to be most important when communicating with their constituents, but find face-to-face meetings, e-mail, and the telephone most important when communicating with each other. The importance of constituent communications correlates with the importance predicted by naturalness theory shown in table 5.1 with a correlation coefficient $r = 0.76$, a relatively high level of correlation. The importance of peer communications correlates with the importance predicted by naturalness theory with a correlation coefficient of $r = 0.63$, a moderate level of correlation. It is expected that the lower correlation between the importance of peer communications and the hypothesized importance shown in table 5.1 is related to the business of legislation. It doesn't matter as much if a communication technology is natural when legislators communicate with each other. This finding is buttressed by legislator comments contained in chapter 4 which indicate that legislators prefer face-to-face interactions whenever possible, but they find e-mail efficient for the "business" aspect of their legislative duties.

FREQUENCY OF USE OF COMMUNICATION TECHNOLOGIES

Table 5.3 highlights the differences in the frequency of use of various communication technologies when legislators communicate with their peers and when legislators communicate with their constituents. The statistical significance shown in the peer mean value column reflects the difference of means between the constituent mean value for communication technology frequency of use and the peer mean value for communication technology frequency of use. For example, the mean value of the importance of hardcopy letters when communicating with constituents is 3.73, somewhere between moderately use weekly and use daily, while the mean peer frequency of use value for hardcopy letters is 1.93, falling somewhere between use annually and use

Table 5.3 Constituent and Peer Communication Technology Frequency of Use Rankings

Communication Technology	*Constituent Frequency of Use Ranking*	*Constituent Mean Value*	*Peer Frequency of Use Ranking*	*Peer Means and Difference of Peer and Constituent Mean t-test Results*
Letters (Hardcopy)	1	3.73	6	1.93***
Telephone	2	3.28	3	3.36***
Face-to-Face Meetings	3	3.04	4	3.28***
Web Page	4	2.49	9	1.18***
Twitter™	5	2.39	8	1.37***
Automated Phone Calls	6	2.15	N/A	N/A
Text Messages	7	1.91	10	0.51***
Facebook™	8	1.45	5	2.32***
Press Releases	9	1.35	N/A	N/A
E-mail	10	1.34	1	3.98***
Television	11	0.89	N/A	N/A
Town-hall Meetings	12	0.50	N/A	N/A
YouTube™	13	0.47	7	1.52***
Radio	14	0.46	N/A	N/A
Blog	15	0.38	2	3.44***

*p< = 0.05, **p< = 0.01, ***p< = 0.001.
Self-Generated by author using Microsoft Word.

monthly. The difference of means between peer and constituent communications with hardcopy letters is statistically significant at the 99.9 percent confidence interval.

Table 5.3 makes it clear that legislators use communication technologies differently depending on the target of the communication, peer or constituent. One of the largest differences is the use of e-mail when communicating with peers and constituents. Legislators use e-mail with their peers at almost four times the frequency that they use e-mail to communicate with constituents, *even though*, as highlighted in the legislator interviews in chapter 4, according to legislators interviewed; constituents communicate with their legislators most frequently via e-mail. This difference suggests a significant fundamental problem associated with constituent communications; there is a mismatch in the way legislators prefer to communicate with constituents and the way that constituents prefer to communicate with legislators. As outlined in chapter 1, communication technology mismatches result in an open loop in the policy feedback process. The more constituent feedback open loops there are in the policy process, the more stable the policy process is. This finding will be discussed in more detail in chapter 8.

COMPARING CONSTITUENT AND PEER COMMUNICATIONS, MATURE TECHNOLOGIES

Based on the findings in the previous section, it is clear that the frequency and use and importance of mature communication technologies vary as a function of the target of the communication: peers or constituents. Kernel density graphs are used to examine the relationships between peer and constituent communication technology frequency of use and importance. Figure 5.1 contains the kernel density plot for peer and constituent mature communication technology frequency of use. Figure 5.2 contains the kernel density plot for peer and constituent mature communication technology importance. Both figure 5.1 and figure 5.2 graph the averages of the three mature communication technologies.

The kernel density plots shown in figures 5.1 and 5.2 highlight the distribution differences between peer and constituent communications. With respect to frequency of use, figure 3.1 shows that peer communications are relatively normal in distribution (skewness = −0.213), while constituent communications are skewed left (−0.480) and more peaked (kurtosis = 3.36) than peer communications (kurtosis = 2.66). These results suggest that with respect to frequency of communications, legislators communicate more frequently with

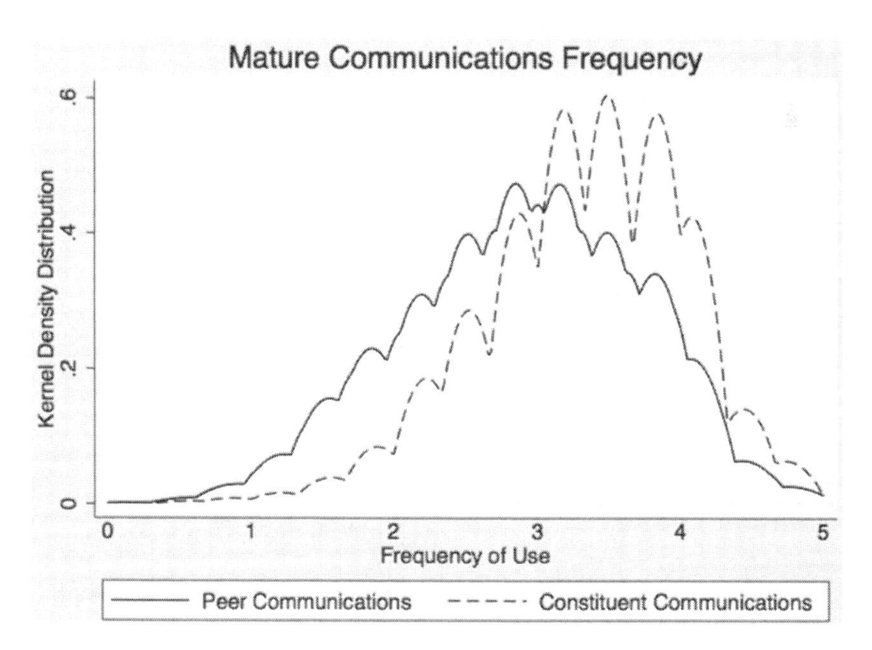

Figure 5.1 Kernel density plot, mature communication technology frequency of use. Self-Generated by author using Stata Quantitative Analyses Program.

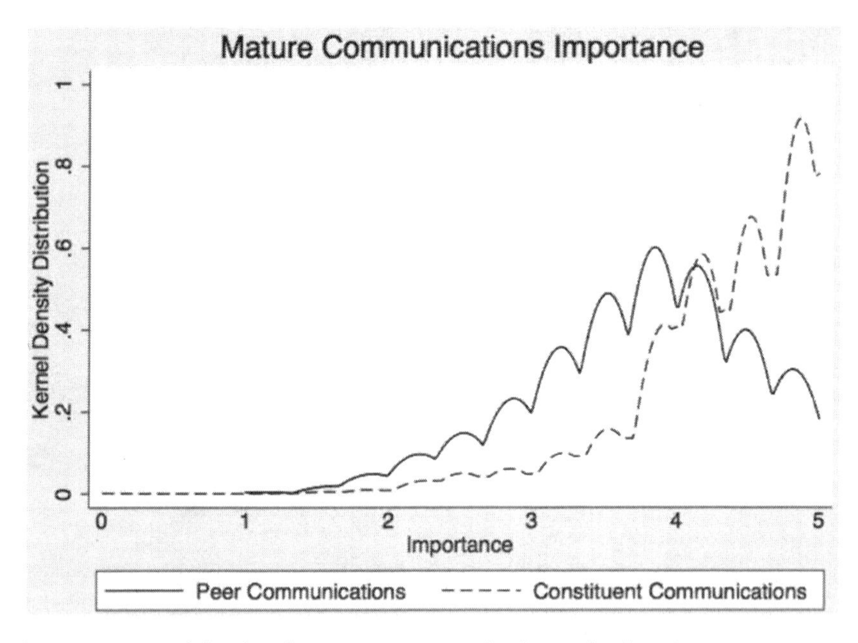

Figure 5.2 Kernel density plot, mature communication technology importance.
Self-Generated by Author using Stata Quantitative Analyses Program.

constituents than they do their peers. Additionally, there is a smaller variance associated with the frequency of use of constituent communications (0.510) than peer communications (0.691). Figure 5.1 suggests that mature communications are more frequently used for constituent communications (left skew) and are also more peaked, with a smaller variance than peer mature communications.

With respect to the importance of mature communications with peers and constituents shown in figure 5.2, a much greater contrast can be seen between peer and constituent communications. The importance of peer mature communications is skewed left (−0.498) less than the importance of constituent mature communications (−1.49). Once again, constituent communications have a smaller variance (0.410) than peer communications (0.581). These results suggest that legislators find constituent communications more important than peer communications. This result is confirmed by the data in table 5.2 where, in all cases, the mean and mode of the importance of constituent communications via mature communication technologies are greater than or equal to the mean and mode of importance of peer communications using mature communication technologies.

To examine the relationships between peer and constituent communications further, the frequency of use and importance of peer and constituent mature communications (in composite forms as the mean of all three mature communication technologies) were examined using Mann-Whitney-Wilcoxon tests and the statistically significant results noted. When comparing the frequency of mature communications with peers versus the frequency of communications with constituents (H_0: *pfreqmature* = *cfreqmature*), the null hypothesis is rejected (z = −20.80). When comparing the importance of mature communications with peers versus the importance of communications with constituents (H_0: *pimportmature* = *cimportmature*), the null hypothesis is once again rejected (z = −32.18). As expected, in *both* cases, constituent mature communication technology frequency of use and importance were higher than peer mature communication technology frequency of use and importance.

COMPARING CONSTITUENT COMMUNICATIONS TO PEER COMMUNICATIONS, IECT

Based on the findings in the previous section, it is clear that the frequency and use and importance of mature communication technologies vary as a function of the target of the communication: peers or constituents. In this section, we examine Internet-enabled communication technologies using kernel density graphs. Kernel density graphs are used to examine the relationships between peer and constituent communication technology frequency of use and importance. Figure 5.3 contains the kernel density plot for peer and constituent mature communication technology frequency of use. Figure 5.4 contains the kernel density plot for peer and constituent mature communication technology importance. Both figure 5.3 and figure 5.4 graph kernel densities of the averages of the seven IECT communication technologies.

The kernel density plots shown in figures 5.3 and 5.4 highlight the distribution differences between peer and constituent communications. With respect to frequency of use, figure 5.3 shows that peer communications are skewed right (skewness = 0.558) and are more peaked than constituent communications (kurtosis = 2.80), while constituent communications are skewed right slightly less (0.484) and are similarly peaked (kurtosis = 2.84) than peer communications. These results suggest that with respect to frequency of communications, legislators communicate more frequently with peers than they do their constituents. Additionally, there is a smaller variance associated with the frequency of use of constituent communications (0.757) than peer communications (0.826). From a practical perspective, the frequency of use of IECT for peer and constituent communications is significantly more similar than the frequency of use of mature communication for peer and constituent technologies.

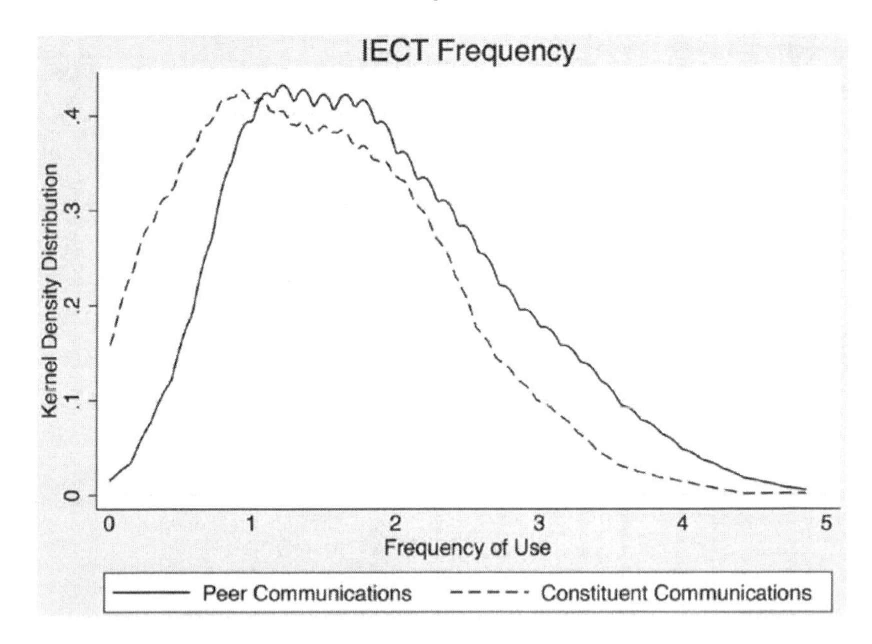

Figure 5.3 Kernel density plot, IECT frequency of use.
Self-generated by author using Stata Quantitative Analyses Program.

With respect to the importance of IECT communications with peers and constituents shown in figure 5.4, a much greater contrast can be seen between peer and constituent communications. The importance of peer IECT communications is skewed right (0.731) more than the importance of constituent IECT communications (0.497). Unlike mature communications, IECT constituent communications have a larger variance (0.989) than peer communications (0.686). Unlike the kernel density plot for IECT frequency of use where peer and constituent communications were peaked about the same, with respect to IECT importance, peer communications are peaked (kurtosis = 3.59) much more than constituent communications (kurtosis = 2.84). These results suggest that legislators find IECT peer communications overall more important than IECT constituent communications. As with the frequency of use of IECT, the importance of IECT is similar whether a legislator is communicating with a peer or a constituent. Mature communication technologies were significantly more important when communicating with constituents than when communicating with other legislators.

To examine the relationships between peer and constituent communications further, the frequency of use and importance of peer and constituent mature communications (in composite forms as the mean of all three mature communication technologies) were examined using Mann-Whitney-Wilcoxon tests

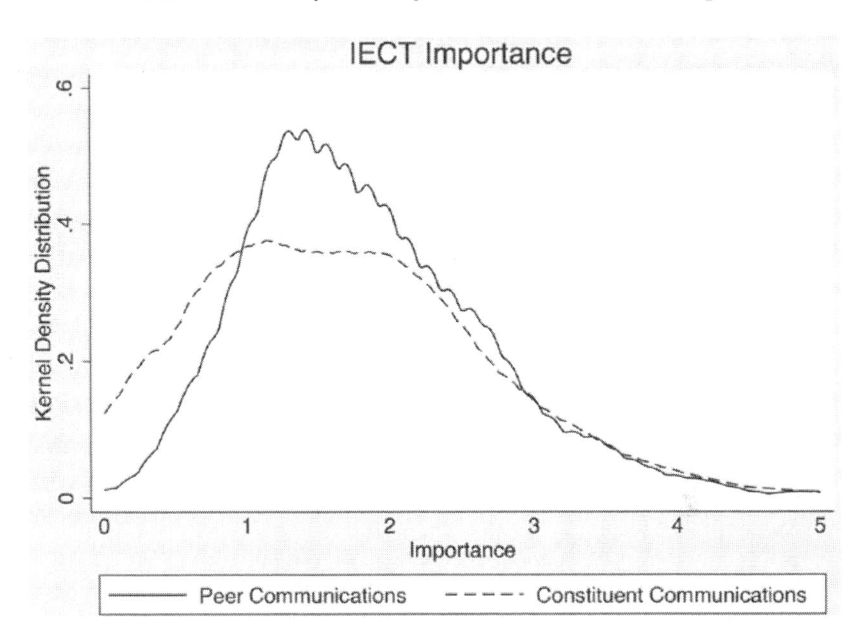

Figure 5.4 Kernel density plot, IECT importance.
Self-generated by author using Stata Quantitative Analyses Program.

and the statistically significant results noted. When comparing the frequency of mature communications with peers versus the frequency of communications with constituents (H_0: *pfreqiect* = *cfreqiect*), the null hypothesis is rejected (z = 24.0). When comparing the importance of Internet-enabled communications with peers versus the importance of Internet-enabled communications with constituents (H_0: *pimportiect* = *cimportiect*), the null hypothesis is once again rejected (z = 8.14). This time, unlike mature communications, both frequency of use and importance of IECT communications were greater with peer communications than with constituent communications.

CONCLUSION

The results in this chapter confirm what a review of the analytical results presented in the previous three chapters imply; legislators use and value (find important) communication technologies differently based on whether they are communicating with each other or with their peers and legislators use and value communication technology differently when using mature communication technology or Internet-enabled communication technology. On average, when using mature communication technologies, legislators prefer and value mature communication technologies when used with constituents, but legislators use and value IECT more when communicating with peers than when communicating with constituents.

From the perspective of the critical frequency theory, it is important to note that just because legislators use and value a particular communication technology to communicate with one category of individuals (such as their constituents), it does not mean that legislators use and value the communication technology the same with another category of individuals (such as other legislators). This finding adds to the complexity of the critical frequency theory and suggests that there may be other categories of individuals (such as administrators or members of the executive branch) for which legislators use and value communication technology differently than when used with constituents or other legislators. This finding suggests that yet another avenue of research is necessary to further develop the critical frequency theory of policy system stability.

NOTE

1. For instance, the symbolic value of a face-to-face meeting to convey bad news might make it a preferred communication channel over a channel with less symbolic value such as e-mail.

Chapter 6

Legislator Roles, Policy Conflict, and Constituent Communications

OVERVIEW

The research in this chapter examines demographic, institutional, and political forces associated with two primary dependent variables: (1) state legislators' perception of their role as a Delegate, Trustee, or Politico and (2) legislators' perception of conflicting constituent policy needs in their districts. Additionally, communication linkages between legislators and constituents are examined in order to understand how they relate to legislator role and legislator perception of district conflict. Legislator–constituent linkages are important because they represent the path by which legislators learn about their constituents' needs, knowledge that is fundamental both to legislator role and legislator perception of constituent policy conflicts in their districts. These communication linkages include measured frequency of use and importance of mature, mass media, and Internet-enabled communication technologies discussed in previous chapters. This chapter deviates slightly from previous chapters insofar as the focus is not specifically on how legislators use and value communication technology as in previous chapters, but rather, this chapter examines the relationships between legislator roles and policy conflict in a legislator's district and communication technology frequency of use and importance.

Legislator role and constituent policy conflicts are examined under three categories of independent variables: (1) political, (2) demographic, and (3) institutional. Political variables include political party affiliation, majority/minority party status, state-level gerrymandering, and ideological polarization among others. Demographic variables include legislator education and age, and institutional variables include legislative chamber,[1] years in office, and legislator state.

LEGISLATIVE ROLES

"It is my aim, to give the people the substance of what I knew they desired, and what I thought was right whether they desired it or not." That Woodrow Wilson would bestow the title of "the best maxim of statesmanship among a free people" (Rohr, 1986, 230) to the words spoken by Edmund Burke above hints at Burke's importance. Indeed, the roots of the legislative role (beyond that of drafting and passing laws) are often attributed to the English conservative who, in the late eighteenth century, had the insight to break legislative roles into two distinct aspects: the *focus* of representation and the *style* of representation. Burke then broke representation focus into two foci for legislators: (1) local interests and (2) national interests. As the opening quote to this section infers, Burke viewed legislators primarily as Trustees and viewed constituents as unenlightened. As we will see later in this chapter, today's legislators see themselves quite differently. The following sections examine the evolution of "Burkean" roles.

Eulau et al. (1959) expanded the Burkean roles to include three variations: (1) the Delegate, (2) the Trustee, and (3) the Politico. Eulau described these roles as follows: (1) the Delegate was a role in which the legislator believes that they are bound by the will of their constituents and must act as an arm of their constituents; (2) the Trustee as a role in which the legislator is bound by their own will and expertise; and (3) the Politico as a hybrid role of the first two in which the legislator is bound to make decisions based on the political situation, sometimes acting as a Trustee and sometimes as a Delegate as the situation warrants. Importantly, Eulau et al. discarded Burke's notion that constituents were unenlightened.

Wahlke (1962) identifies a legislator role he identifies as the "areal role" (13). The areal role (one of several clientele roles Wahlke identifies) is a legislator orientation that focuses on the needs of a district rather than the needs of a state. In the same book that Wahlke suggests that legislators will act more like Trustees, he notes that legislators who identify with the areal role can be expected to act more like a Delegate than a Trustee. Effectively, Wahlke argues that depending on how legislators see themselves (state focused or district focused), they may identify as either a Trustee or a Delegate. Countering Wahlke's state-focused research, a study by Carey et al. (2006) suggests that term limits will force legislators toward the Trustee role advocated by Edmund Burke (1891). Carey et al.'s theory that term limits will force legislators toward the role of a Trustee is confirmed by the research presented in this book. Legislators from states with term limits are 3 percent more likely to identify as a Trustee than legislators from states that do not have term limits ($p = 0.10$). While several studies suggest that legislators tend to view themselves as Trustees (Richardson and Cooper, 2006; Wahlke, 1962), a finding contradicted by the results outlined in this chapter; legislators are 4.5 percent more likely to identify as a Delegate than a Trustee.

Although there are three theoretical legislator roles, from a practical perspective, these three roles can be thought of as a single role: the Politico. If one thinks of legislator roles as a continuum, on one end of the continuum is the Trustee, a legislator who makes decisions solely based on his/her own will and expertise. At the other end of the continuum lies the Delegate. A pure Delegate responds entirely to the *perceived* will of their constituents. Lying somewhere between the Trustee and the Delegate is the Politico. For the purposes of this book, the continuum will be called the Politico scale.

Although the legislative roles of Delegate, Trustee, and Politico are widely accepted by scholars, they are not universally accepted. In *Rethinking Representation*, Mansbridge (2003) extends earlier work on legislator roles by presenting four forms of representation: (1) Promissory, (2) Anticipatory, (3) Gyroscopic, and (4) Surrogate. Two of the four forms of representation are associated with the use of communication technology to communicate with their constituents (promissory, anticipatory, and surrogate), while one, gyroscopic representation, much like the ideal Trustee, does not depend on constituent communications. Under promissory representation, a voter elects a legislator based on what the legislator indicates that they will do for the voter in office (the promise) and communicates the completion of these promises to the voter. The voter evaluates the legislator's performance (again, based on communications, not only from the legislator, but from the media, special interest groups, and other stakeholders) and then rewards or punishes the legislator during reelection.

Anticipatory representation reflects a legislator's efforts to please voters during the next election cycle rather than the voters who put them in office (although clearly there can be a great deal of overlap). Anticipatory legislators try to determine, and communicate with, voters who are likely to support them during their next election. Surrogate representatives represent citizens in special topic areas not necessarily associated with *their* constituents. For example, former US Representative Barney Frank was a surrogate representative for lesbian, gay, bisexual, transgender, and queer (LGBTQ) citizens. Frank, a homosexual man himself, used his political office to further LGBTQ legislation, even when the legislation may not have directly impacted his constituents. In effect, Frank became a surrogate representative for the LGBTQ community. Gyroscopic representation reflects legislators who act for "internal reasons" (520); gyroscopic representatives are similar to the Trustee representative in that communications with constituents are not necessary. Importantly, while Mansbridge's work is noted here due to its significance in the literature, her legislator roles were not instrumented, and instead, the survey instrument focused on measuring the Politico scale.

Because researchers suggest that location on the Politico scale will vary as a function of the exact policy in question (Arnold, 1992; Eulau et al., 1959; Oleszek, 2011), this variable is treated as an average of all policy positions that the legislator has experienced[2]. Varying legislator roles as a function of the exact policy in question can be problematic; a survey instrument would need to be created that examined each significant policy in each legislator's district. Such an instrument would need to be customized for every legislative district in every state and would be a significant undertaking far outside the capacity of a single researcher working alone.

Figure 6.1 highlights legislator responses to the question: In the event your position on an issue differs from the position of the majority of your constituents, approximately what percentage of the time would you say that you choose to vote the position of the majority of your constituents instead of the position you personally feel is best? This question forms the foundation for the Politico scale.

The Politico scale variable was utilized in bivariate regressions to test for statistically significant differences of means. The results of these bivariate regressions are discussed in the sections below. In addition to the Politico scale variable that varies from 0 to 100, a Politico scale binary variable was created. If a legislator's self-perceived Politico scale number was <50, the Politico scale binary was coded as a 1. If a legislator's self-perceived Politico

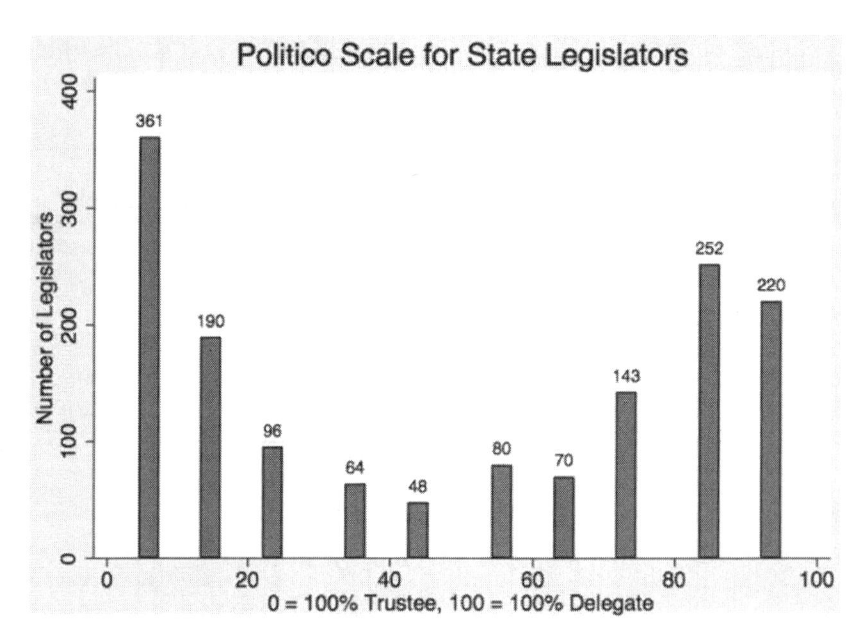

Figure 6.1 State legislator politico scale.
Self-generated by author using Stata Quantitative Analyses Program.

scale number was ≥50, the Politico scale binary was coded as a 0. With respect to the Politico scale binary variable, only one measure, education, was statistically significant. Evaluated at the mean value of education (17.3 years of education), the marginal effects of education on the Politico binary variable suggests that a one-year increase in legislator education is associated with a 1.6 percent in a legislator acting as a Trustee, ceteris paribus ($z < .001$).

With respect to the ordinal Politico (non-binary) scale, bivariate regression results suggest that of all the political, demographic, and institutional variables analyzed, only education, race, and Senate membership were statistically significant. The education result indicates that for every year increase in legislator education, a legislator is likely to act 1.22 percent more as a Trustee, on average, all else equal. The Senate result indicates that Senators are 3.79 percent more likely to act as Trustees than are Representatives. The race results indicate that black legislators are 11.4 percent more likely to identify as Delegates than white legislators.

State-level fixed effects analyses on the Politico scale suggest that Delaware legislators act 23 percent more like Delegates than do Alabama legislators. Kansas legislators act 18.1 percent more like Delegates than do Alabama legislators. Minnesota legislators act 21.7 percent more like Delegates than do Alabama legislators. Oregon legislators act 23.7 percent more like Delegates than do Alabama legislators. Rhode Island legislators act 15.1 percent more like Delegates than do Alabama legislators. And finally, Wisconsin legislators act 19 percent more like Delegates than do Alabama legislators. These state-level fixed effects explain relatively little of the variation in legislator role with an r-squared of 0.92 percent. Figure 6.2 highlights the Politico scale for each state with a legislator that responded to the survey.

Bivariate regressions were completed on each form of communication technology and the Politico scale. Both communication technology frequency of use and communication technology importance are tested with respect to peer communications and constituent communications.

With respect to the Politico scale and constituent communication *frequency*, the results suggest that all statistically significant communication technologies when used to communicate with constituents move a legislator toward the role of a Delegate. The most significant of these communication technologies is telephone use, followed by mass-media radio, mass-media press, face-to-face meetings, mass-media television, hardcopy letters, web pages, and town-hall meetings. Using the telephone results as an example, a one category increase in telephone frequency of use with constituents is correlated with a 4.74 percent increase in a legislator feeling like a Delegate, ceteris paribus. These results suggest that the more a legislator communicates with a constituent, the more they feel like a Delegate. These results are hardly surprising.

State	4.5	14.5	24.5	34.5	44.5	54.5	64.5	74.5	84.5	90-100%	Total
	When in Conflict, % Time a Legislator Chooses Constituent Will Over Own Will										
Alabama	8	4	3	2	2	1	1	1	3	4	29
Alaska	5	2	3	1	0	1	1	1	0	3	17
Arizona	7	3	0	1	1	0	0	1	1	6	20
Arkansas	9	4	0	1	1	3	0	0	10	4	32
California	1	1	3	1	2	1	0	0	2	0	11
Colorado	6	4	1	1	0	0	2	1	3	2	20
Connecticut	7	1	1	2	0	1	3	2	3	2	22
Delaware	2	2	0	0	0	0	1	3	5	3	16
Florida	7	2	0	1	0	0	1	3	2	3	19
Georgia	9	7	5	2	2	4	1	2	7	7	46
Hawaii	7	3	7	1	0	0	3	5	0	4	30
Idaho	13	6	0	0	1	1	3	2	5	3	34
Illinois	5	3	2	0	1	1	0	3	5	5	25
Indiana	3	1	0	0	1	0	2	1	4	2	14
Iowa	9	2	2	0	4	0	4	4	9	4	38
Kansas	9	1	3	1	0	1	1	6	11	8	41
Kentucky	3	2	0	3	0	0	0	1	3	3	15
Louisiana	5	3	1	0	0	3	2	5	8	8	35
Maine	16	8	2	1	0	5	3	7	6	7	55
Maryland	15	5	1	1	2	1	1	4	12	8	50
Massachusetts	13	9	1	6	1	0	0	1	4	5	40
Michigan	2	2	2	1	1	1	4	4	2	3	22
Minnesota	8	2	1	0	0	0	1	1	3	3	19
Mississippi	1	1	1	1	0	0	1	2	2	5	14
Missouri	12	2	2	1	0	3	0	1	5	9	35
Montana	10	6	1	0	1	2	1	7	4	4	36
Nebraska	2	4	1	0	2	5	1	2	6	1	24
Nevada	1	1	1	1	1	0	0	1	1	0	7
New Hampshire	25	10	10	5	5	15	11	16	20	16	133
New Jersey	0	4	0	0	1	0	1	0	3	0	9
New Mexico	5	0	3	2	1	3	0	1	1	5	21
New York	10	6	0	0	0	1	1	0	4	4	26
North Carolina	14	9	6	2	1	5	0	6	5	10	58
North Dakota	18	7	1	3	0	3	1	6	14	0	53
Ohio	2	2	0	1	0	0	0	0	2	3	10
Oklahoma	8	2	2	0	1	1	1	6	5	2	28
Oregon	1	4	3	4	1	1	0	1	1	1	17
Pennsylvania	4	5	1	2	2	4	1	3	5	8	35
Rhode Island	3	2	3	0	1	2	2	5	4	6	28
South Carolina	3	4	2	1	0	0	0	0	3	2	15
South Dakota	9	2	0	1	0	1	0	3	9	2	27
Tennessee	0	4	2	5	2	0	0	2	1	5	21
Texas	3	2	1	0	1	0	0	3	4	4	18
Utah	11	5	3	2	3	1	1	1	7	5	39
Vermont	14	7	5	1	2	1	4	6	12	8	60
Virginia	5	7	1	1	1	0	0	0	1	3	19
Washington	6	4	2	2	0	1	1	2	2	6	26
West Virginia	9	4	1	0	2	2	4	5	4	3	34
Wisconsin	2	2	1	0	1	1	1	1	5	4	18
Wyoming	7	4	2	3	0	2	1	4	8	4	35
Total	354	187	93	64	48	78	67	142	246	217	1,496

Figure 6.2 Politico scale by state.
Self-generated by author using Stata Quantitative Analysis Program.

With respect to the Politico scale and the *importance* of communicating with constituents, the results suggest that the importance of all statistically significant communication technologies when used to communicate with constituents move a legislator toward the role of a Delegate. The most significant of these communication technologies is face-to-face meetings with constituents, followed by telephone, text messages, mass-media radio, hardcopy letters, blogs, and web pages. Using the face-to-face results as an example, a one category increase in the importance of face-to-face meetings with constituents is correlated with a 2.88 percent increase in a legislator feeling like a Delegate, ceteris paribus.

Basic, Intermediate, and Advanced Regression Models

Three models were created to determine the relationships between the politico score and DPI variables. The basic model contains legislator demographic information, the intermediate model adds primary political and institutional variables, and the advanced model adds secondary institutional variables. Additionally, the results of these three models are not presented in a table, but rather, are discussed in the paragraphs following the models presented below:

Basic Model: $politico = \beta_0 + \beta_1 (age) + \beta_2 (race) + \beta_3 (male) + \beta_4 (education) + \mathcal{E}$.
Intermediate Model: $politico = \beta_0 + \beta_1 (age) + \beta_2 (race) + \beta_3 (male) + \beta_4 (education) + \beta_5 (conservativism) + \beta_6 (minority\ status) + \beta_7 (years\ in\ office) + \beta_8 (senate) + \mathcal{E}$.
Full Model: $politico = \beta_0 + \beta_1 (age) + \beta_2 (race) + \beta_3 (male) + \beta_4 (education) + \beta_5 (conservativism) + \beta_6 (minority\ status) + \beta_7 (years\ in\ office) + \beta_8 (senate) + \beta_9 (gerrymandering) + \beta_{10} (professional\ legislature\ binary) + \beta_{11} (term\ limits) + \beta_{12} (conflict\ scale) + \beta_{13} (polarization\ binary) + \mathcal{E}$.

The basic model is statistically significant for race (unstandardized coefficient = 13.1, $p \leq 0.01$) and education (unstandardized coefficient = −1.4, $p \leq 0.001$). With respect to race, compared to white legislators, black legislators are 13.1 percent more likely to identify as a delegate, on average, all else equal. With respect to education, a one-year increase in legislator education is associated with a 1.4 percent increase in the likelihood that the legislator will identify as a trustee, on average, all else equal.

The intermediate model is statistically significant for race (unstandardized coefficient = 13.23, $p \leq 0.01$) and education (unstandardized coefficient = −1.3, $p \leq 0.001$). With respect to race, compared to white legislators, black legislators are 13.23 percent more likely to identify as a delegate, on average, all else equal. With respect to education, a one-year increase in legislator education is associated with a 1.3 percent increase in the likelihood that the legislator will identify as a trustee, on average, all else equal. Importantly, the addition of political variables changed the statistically significantly coefficients very little; however, the adjusted r-squared decreased in the intermediate model suggesting that the additional variables offer no explanatory power over the basic model.

The full model is statistically significant for race (unstandardized coefficient = 11.6, $p \leq 0.01$), education (unstandardized coefficient = −1.2, $p \leq 0.01$), and the conflict scale (unstandardized coefficient = 0.41, $p \leq 0.001$). With respect to race, compared to white legislators, black legislators are 11.6 percent more likely to identify as a delegate, on average, all else equal. With respect to education, a one-year increase in legislator education is associated with a 1.2 percent increase in the likelihood that the legislator will identify as a trustee, on average, all else equal. With respect to the conflict scale, a 1 percent increase in the amount of policy conflict detected by a legislator in their district is associated with a 0.4 percent increase in the likelihood that the legislator will identify as a delegate, on average, all else equal. To put this last result in common terms, the more policy conflict in a legislator's district, the more likely they are to act like a delegate. Unlike the results for the intermediate model, the full model adjusted r-squared value increased from 1.6 percent to 4.5 percent, a significant change in the explanatory power of the model.

Full Regression Models

The previous sections analyzed and discussed three regression models with politico behavior as the dependent variable. In this section, a legislator's politico score is examined as a function of communication technology frequency of use and importance.

$$politico = \beta_0 + \beta_1(pfreqmature) + \beta_2(pimportmature) + \beta_3(pfreqiect) + \beta_4(pimportiect) + \beta_5(cfreqmature) + \beta_6(cimportmature) + \beta_7(cfreqmass) + \beta_8(cimportmass) + \beta_9(cfreqiect) + \beta_{10}(cimportiect) + \mathcal{E}.$$

In this model, the only communication technology frequency of use and importance that has a statistically significant relationship with trustee/delegate behavior is the frequency of use of mature communication technologies (unstandardized coefficient = 4.8, $p \leq 0.01$). A one unit increase in the frequency of use of a mature communication technology to communicate with constituents is associated with a 4.8 percent increase in a legislator behaving as a delegate, on average, all else equal.

A second regression model, identical to the above model with the exception of the addition of all DPI variables:

$$politico = \beta_0 + \beta_1(pfreqmature) + \beta_2(pimportmature) + \beta_3(pfreqiect) + \beta_4(pimportiect) + \beta_5(cfreqmature) + \beta_6(cimportmature) + \beta_7(cfreqmass) + \beta_8(cimportmass) + \beta_9(cfreqiect) + \beta_{10}(cimportiect) + \beta_{11}(age) + \beta_{12}(race) + \beta_{13}(male) + \beta_{14}(education) + \beta_{15}(conservativism) + \beta_{16}(minority\ status) + \beta_{17}(years\ in\ office) + \beta_{18}(senate) + \beta_{19}(gerrymandering) + \beta_{20}(professional\ legislature\ binary) + \beta_{21}(term\ limits) + \beta_{22}(conflict\ scale) + \beta_{23}(polarization\ binary) + \mathcal{E}.$$

In this model, the importance of mature communication technologies when used with peers (unstandardized coefficient = −3.7, $p \leq 0.05$), the frequency of use of mature communication technologies when used with constituents (unstandardized coefficient = 6.4, $p \leq 0.01$), the conflict scale (unstandardized coefficient = 0.42, $p \leq 0.001$), and the polarization binary variable (unstandardized coefficient = −7.2, $p \leq 0.05$) are all statistically significant. Putting these results in a more digestible format, increases in the importance of mature communications with peers and legislator polarization behaviors are both associated with a legislator acting as a Trustee, while increases in legislator detection of district policy conflict and increases in the frequency of use of mature communication technologies when communicating with constituents are both associated with legislators acting as Delegates.

DISTRICT POLICY CONFLICT

According to Arnold (1992), legislators balance the conflicting requirements of constituents by examining the costs and benefits of legislation on their constituents. Legislators are averse to legislation that imposes costs on their constituents and enjoy passing legislation that benefits their constituents with minimal or no costs (1992, 4). Because legislators are driven by the need to be reelected (Mayhew, 1974), they attempt to *anticipate* (Kingdon and Thurber, 2003, 73) constituent reactions toward a particular piece of legislation and will vote in a manner they believe will precipitate their greatest chance for reelection. In an opposing argument, Miller and Stokes (1963) found that the links between a legislators roll-call vote and their constituents policy positions were weak. Arnold (1992, 9) notes that some scholars have concluded that legislators need not be concerned with their constituents needs at all because in general, constituents are poorly informed about how their legislators vote on policy issues. Other early research citing the Miller-Stokes theory (Arnold, 1992; Erikson, 1978; Hedlund and Friesema, 1972; Jewell, 1983) suggests that little evidence existed to refute the weak links that Miller and Stokes found between legislators roll-call voting behavior and the policy positions of their constituents.

The job of divining the preferences of their constituents when facing conflicting needs is made somewhat easier because legislators typically decide between paired alternative choices (Arnold, 1992; Kahneman and Tversky, 1979). Legislators, they argue, need to decide whether either of the two alternatives in a yay/nay vote is more likely to arouse the ire of their constituents. In addition to deciding on paired alternative choices, legislators who are unwilling to commit to a decision may avoid taking a stance on a legislative issue by not voting on the issue at all. Arnold (1992) argues that not only legislators need to estimate the preferences of their constituents, but, equally, they need to estimate whether or not constituents who disagree with their position (another source of conflict for legislators) might impact their choices in congressional elections. According to Arnold, a legislator's fear is not that there are citizens who disagree with their vote but rather, that those citizens might disagree strongly enough to mount an opposition campaign or vote for an opponent in elections.

Figure 6.3 highlights the legislator estimated frequency of policy conflicts between their policy preferences and the policy preferences of the majority of their constituents and figure 6.4 lists the percentage of time that legislators estimate that their policy preferences conflict with the policy preferences of their constituents by state.

As shown in figure 6.3, a vast majority of legislators estimate that policy conflicts between their policy preferences and the policy preferences of constituents in their districts occur infrequently.

As with the Politico scale analyzed in the previous section, bivariate regressions were completed using legislator perception of district conflict as the dependent variable and DPI variables. The results of these bivariate regressions show that of all the political, demographic, and institutional variables analyzed, only the Republican/Democrat binary variable, polarization score, legislator age, legislator education, and state-level fixed effects had statistically significant relationships with district policy conflict estimations by legislators. The Republican result indicates that Republicans report 1.58 percent less constituent policy conflict than Democrats, ceteris paribus. The age coefficient suggests that for every one year of legislator age, legislators report .11 percent more constituent policy conflict, ceteris paribus. The education result indicates that for every year increase in legislator education, a

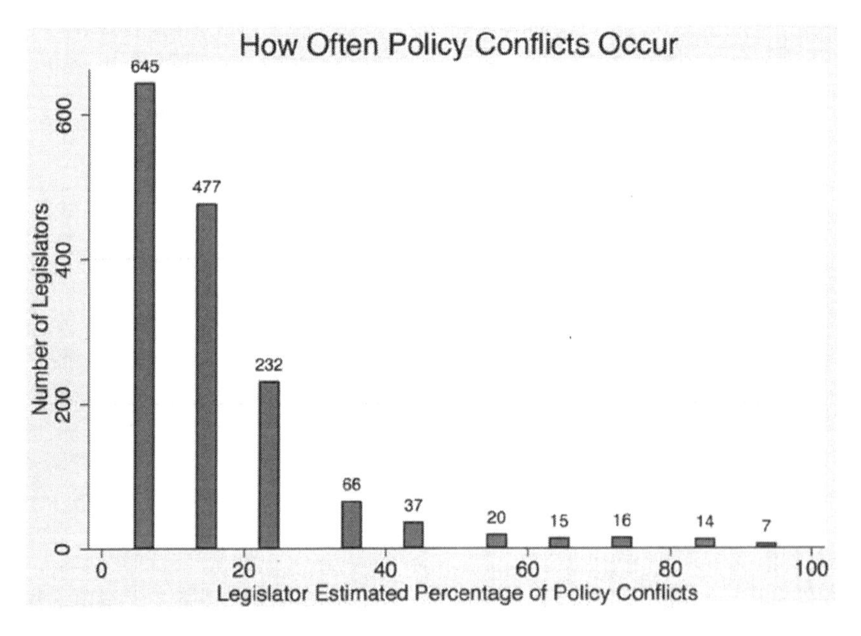

Figure 6.3 State legislator district conflict.
Self-generated by author using Stata Quantitative Analyses Program.

State	Percent of Time a Legislator's Policy Preferences Conflict With Their Constituen										Total
	4.5	14.5	24.5	34.5	44.5	54.5	64.5	74.5	84.5	90-100%	
Alabama	11	7	7	1	2	0	0	0	1	0	29
Alaska	13	5	0	1	0	0	0	0	0	0	19
Arizona	11	6	1	0	1	0	0	0	1	0	20
Arkansas	18	8	3	0	1	0	1	0	0	0	31
California	6	2	3	0	0	0	0	0	0	0	11
Colorado	8	12	2	0	0	0	0	0	0	0	22
Connecticut	11	7	4	0	0	0	0	0	0	0	22
Delaware	6	3	6	1	0	0	0	0	0	0	16
Florida	9	3	2	2	0	1	0	1	0	0	18
Georgia	17	13	9	3	1	1	1	1	0	0	46
Hawaii	12	3	11	0	1	1	2	0	0	0	30
Idaho	18	7	2	4	1	1	0	0	1	0	34
Illinois	12	12	1	0	1	0	0	0	0	0	26
Indiana	6	3	4	0	1	0	0	0	0	0	14
Iowa	16	14	4	1	1	0	0	1	0	1	38
Kansas	19	9	5	2	1	1	1	1	0	0	39
Kentucky	8	4	1	0	0	0	0	1	1	0	15
Louisiana	11	16	4	2	2	0	0	0	0	0	35
Maine	23	15	6	6	1	3	0	0	1	0	55
Maryland	21	17	9	2	0	1	0	0	0	1	51
Massachusetts	16	12	8	2	0	0	0	0	1	1	40
Michigan	10	10	1	1	0	0	0	0	0	0	22
Minnesota	12	3	3	0	0	0	0	1	0	0	19
Mississippi	6	6	0	1	0	0	0	0	0	0	13
Missouri	24	9	1	0	0	0	1	0	0	1	36
Montana	16	12	5	2	0	1	1	0	0	0	37
Nebraska	8	8	1	3	1	1	0	1	1	0	24
Nevada	4	3	0	0	0	0	0	0	0	0	7
New Hampshire	44	41	23	7	8	4	2	1	1	1	132
New Jersey	4	3	1	0	1	0	0	0	0	0	9
New Mexico	9	5	6	1	0	0	0	0	0	0	21
New York	14	7	3	0	0	0	0	1	1	0	26
North Carolina	24	18	10	4	1	0	1	0	0	0	58
North Dakota	18	22	9	4	0	0	0	0	0	0	53
Ohio	7	3	1	1	0	0	0	0	0	0	12
Oklahoma	12	8	8	0	0	0	0	0	0	0	28
Oregon	5	4	6	1	1	0	0	0	0	0	17
Pennsylvania	12	11	7	2	1	0	0	0	1	1	35
Rhode Island	8	9	6	1	2	0	0	1	1	0	28
South Carolina	7	6	1	1	0	0	0	0	0	0	15
South Dakota	4	14	3	0	1	0	1	2	1	0	26
Tennessee	8	8	5	0	0	0	0	0	0	0	21
Texas	6	9	1	1	0	0	1	0	0	0	18
Utah	18	16	3	0	0	0	1	1	0	0	39
Vermont	28	11	9	2	4	1	2	1	1	0	59
Virginia	10	7	0	0	0	0	0	1	0	0	18
Washington	11	5	6	4	0	0	0	0	0	0	26
West Virginia	13	14	3	1	0	2	0	1	0	0	34
Wisconsin	11	6	2	0	0	0	0	0	0	0	19
Wyoming	13	16	4	1	1	0	0	0	0	1	36
Total	638	472	220	65	35	18	15	16	13	7	1,499

Figure 6.4 Legislator conflict estimations by state.
Self-generated by author using Stata Quantitative Analysis Program.

legislator reports −.366 less constituent policy conflict, ceteris paribus. The state-level fixed effects suggest that Montana legislators report 8.1 percent less constituent policy conflict than do Alabama legislators. South Dakota legislators report 10.2 percent less constituent policy conflict than do Alabama legislators. Legislator race is not correlated with changes in constituent policy conflict perception.

Bivariate regression results show that of all the political, demographic, and institutional variables analyzed, only the two polarization variables and two state variables were statistically significant. For each one unit increase in a legislator's polarization, there is an associated 1.66 percent decrease in the amount of conflict a legislator detects in their district. When legislators were defined as either polarized or not polarized (the binary polarization measure), legislator polarization was associated with a 3.22 percent reduction in the amount of conflict a legislator detects in their district, on average, all else equal. With respect to the state variables, Montana legislators are 8.1 percent more likely to detect conflict in their district compared to Alabama, and South Dakota legislators are 10.19 percent more likely to detect conflict in their district compared to Alabama legislators.

Bivariate regressions were completed on each form of communication technology and the constituent policy conflict scale. Both communication technology frequency of use and communication technology importance are tested with respect to peer communications and constituent communications.

With respect to the conflict scale and constituent communication *frequency*, the results suggest that some statistically significant communication technologies when used to communicate with constituents increase legislator perception of conflict, while other communication technologies decrease legislator perception of conflict. For example, a one category increase in the frequency of communication with constituents via mass-media radio results in an increase in legislator perception of constituent policy conflict. A one category increase in the use of hardcopy letters to communicate with constituents is associated with a 0.925 decrease in legislator perception of constituent policy conflict, ceteris paribus. Once again, these results are not surprising. As legislators communicate more with their constituents, their sense of conflict decreases. These results also suggest that legislators may communicate more with constituents who agree with their policy positions than with constituents who disagree with their policy positions.

With respect to the conflict scale and the *importance* of communicating with constituents, the results suggest that the importance of all statistically significant communication technologies when used to communicate with constituents move a legislator toward a decreasing sense of constituent policy conflict. For example, a one category increase in the importance of face-to-face meetings with constituents is correlated with a 1.82 percent decrease in legislator perception of constituent policy conflict, ceteris paribus.

Basic, Intermediate, and Advanced Regression Models

Three models were created to determine the relationships between the district conflict score and DPI variables. The basic model contains legislator demographic information, the intermediate model adds primary political and institutional variables, and the advanced model adds secondary institutional variables. Additionally, the results of these three models are not presented in a table, but rather, are discussed in the paragraphs following the models presented below:

Basic Model: $conflict = \beta_0 + \beta_1 (age) + \beta_2 (race) + \beta_3 (male) + \beta_4 (education) + \mathcal{E}.$

Intermediate Model: $conflict = \beta_0 + \beta_1 (age) + \beta_2 (race) + \beta_3 (male) + \beta_4 (education) + \beta_5 (conservativism) + \beta_6 (minority\ status) + \beta_7 (years\ in\ office) + \beta_8 (senate) + \mathcal{E}.$

Full Model: $conflict = \beta_0 + \beta_1 (age) + \beta_2 (race) + \beta_3 (male) + \beta_4 (education) + \beta_5 (conservativism) + \beta_6 (minority\ status) + \beta_7 (years\ in\ office) + \beta_8 (senate) + \beta_9 (gerrymandering) + \beta_{10} (professional\ legislature\ binary) + \beta_{11} (term\ limits) + \beta_{12} (conflict\ scale) + \beta_{13} (polarization\ binary) + \mathcal{E}.$

The basic model is statistically significant for age (unstandardized coefficient = 0.11, $p \leq 0.001$), race (black, unstandardized coefficient = 4.0, $p \leq 0.05$), and education (unstandardized coefficient = −.46, $p \leq 0.01$). The age coefficient indicates that each one additional year in age is associated with a legislator detecting .11 percent more policy conflict in their district, on average, all else equal. With respect to race, compared to white legislators, black legislators detect 4 percent more policy conflict in their district, on average, all else equal. A one-year increase in legislator education is associated with a −.46 percent decrease in the detection of policy conflict in their district, on average, all else equal. The basic model adjusted r-squared value is 1.4 percent

The intermediate model is statistically significant for age (unstandardized coefficient = −0.16, $p \leq 0.001$), education (unstandardized coefficient = −0.48, $p \leq 0.01$), conservativism (unstandardized coefficient = −0.40, $p \leq 0.01$), and years in office (unstandardized coefficient = −0.15, $p \leq 0.05$). With respect to age, a one-year increase in legislator age is associated with a 0.165 percentage increase in the amount of policy conflict detected in a legislator's district. A one-year increase in education is associated with a −0.48 percent decrease in the amount of policy conflict detected in a legislator's district. A one unit increase in conservativism is associated with a 0.4 percent decrease in the amount of policy conflict detected in a legislator's district. A one-year increase in the time a legislator has been in office is associated with a 0.15 percentage decrease in the amount of policy conflict detected in a legislator's district, on average, all else equal. The intermediate model adjusted r-squared value is 1.98 percent.

The full model is statistically significant for age (unstandardized coefficient = 0.19, $p \leq 0.001$), education (unstandardized coefficient = −0.39, $p \leq 0.001$), conservativism (unstandardized coefficient = −0.42, $p \leq 0.05$), years in office (unstandardized coefficient = −0.15, $p \leq 0.05$), politico behavior (unstandardized coefficient = 0.08, $p \leq 0.001$), and polarization behavior (unstandardized coefficient = −2.66, $p \leq 0.01$). With respect to age, a one-year increase in legislator age is associated with a 0.19 percentage increase in the amount of policy conflict detected in a legislator's district. A one-year increase in education is associated with a −0.39 percent decrease in the amount of policy conflict detected in a legislator's district. A one unit increase in conservativism is associated with a 0.42 percent decrease in the amount of policy conflict detected in a legislator's district. A one-year increase in the time a legislator has been in office is associated with a 0.15 percentage decrease in the amount of policy conflict detected in a legislator's district, on average, all else equal. A one unit increase in a legislator's behavior as a Delegate is associated with a 0.08 percentage increase in the amount of policy conflict detected in a legislator's district, on average, all else equal. A legislator self-identifying as strongly conservative or strongly progressive (polarization) is associated with a 2.66 percentage decrease in the amount of policy conflict detected in a legislator's district, on average, all else equal. The full model adjusted r-squared value is 6.2 percent.

Full Regression Models

The previous sections analyzed and discussed three multivariate regression models associated with DPI variables. In this section, a legislator's delegate score is examined as a function of communication technology frequency of use and importance.

$$conflict = \beta_0 + \beta_1(pfreqmature) + \beta_2(pimportmature) + \beta_3(pfreqiect) + \beta_4(pimportiect) + \beta_5(cfreqmature) + \beta_6(cimportmature) + \beta_7(cfreqmass) + \beta_8(cimportmass) + \beta_9(cfreqiect) + \beta_{10}(cimportiect) + \mathcal{E}.$$

In this model, the communication technologies with statistically significant relationships with district conflict detection are the importance of mature communication technologies (unstandardized coefficient = −1.5, $p \leq 0.05$), and the frequency of use of mass media to communicate with constituents (unstandardized coefficient = 4.13, $p \leq 0.01$). A one unit increase in the importance of a mature communication technology to communicate with constituents is associated with a 1.5 percent decrease in a legislator's detection of policy conflict in their district, on average, all else equal. A one unit increase in the frequency of use of a mass-media communication technology to communicate with constituents is associated with a 4.1 percent increase in a legislator's detection of policy conflict in their district, on average, all else equal.

A second regression model, identical to the above model with the exception of the addition of all DPI variables:

$$conflict = \beta_0 + \beta_1(pfreqmature) + \beta_2(pimportmature) + \beta_3(pfreqiect) + \beta_4(pimportiect) + \beta_5(cfreqmature) + \beta_6(cimportmature) + \beta_7$$

(cfreqmass) $+ \beta_8$*(cimportmass)* $+ \beta_9$*(cfreqiect)* $+ \beta_{10}$*(cimportiect)* $+ \beta_{11}$ *(age)* $+\beta_{12}$*(race)* $+\beta_{13}$*(male)* $+\beta_{14}$*(education)* $+\beta_{15}$*(conservativism)* $+\beta_{16}$ *(minority status)* $+ \beta_{17}$*(years in office)* $+ \beta_{18}$*(senate)* $+ \beta_{19}$*(gerrymandering)* $+\beta_{20}$*(professional legislature binary)* $+\beta_{21}$*(term limits)* $+\beta_{22}$*(conflict scale)* $+ \beta_{23}$*(polarization binary)* $+ \mathcal{E}$.

In this model, statistically significant results were found for age (unstandardized coefficient = 0.19, $p \leq 0.001$), gender (unstandardized coefficient = −0.39, $p \leq 0.001$), education (unstandardized coefficient = −0.39, $p \leq 0.001$), gerrymandering (unstandardized coefficient = −0.42, $p \leq 0.05$), years in office (unstandardized coefficient = −0.15, $p \leq 0.05$), politico behavior (unstandardized coefficient = 0.08, $p \leq 0.001$), polarization behavior (unstandardized coefficient = −2.66, $p \leq 0.01$), the importance of mature communication technologies when used with peers (unstandardized coefficient = −2.1, $p \leq 0.05$), the frequency of use of mature communication technologies when used with constituents (unstandardized coefficient = −2.5, $p \leq 0.01$), the frequency of use of mass media when used to communicate with constituents (unstandardized coefficient = 4.2, $p \leq 0.01$), and the importance of mass media when used to communicate with constituents (unstandardized coefficient = −2.5, $p \leq 0.05$). With respect to age, a one-year increase in legislator age is associated with a 0.19 percentage increase in the amount of policy conflict detected in a legislator's district. A one-year increase in education is associated with a −0.39 percent decrease in the amount of policy conflict detected in a legislator's district. A one unit increase in conservativism is associated with a 0.42 percent decrease in the amount of policy conflict detected in a legislator's district. A one-year increase in the time a legislator has been in office is associated with a 0.15 percentage decrease in the amount of policy conflict detected in a legislator's district, on average, all else equal. A one unit increase in a legislator's behavior as a Delegate is associated with a 0.08 percentage increase in the amount of policy conflict detected in a legislator's district, on average, all else equal. A legislator self-identifying as strongly conservative or strongly progressive (polarization) is associated with a 2.66 percentage decrease in the amount of policy conflict detected in a legislator's district, on average, all else equal. The communication technology relationships can be interpreted exactly the same as the previous model and will not be reinterpreted here. The full model adjusted r-squared value is 8.3 percent.

CONCLUSION

State legislators view themselves mostly as Delegates slightly more than Trustees with a mean Politico scale score of 54.5. More importantly, with a mean conflict scale score of 14.5, legislators perceive relatively little policy conflict among the constituents in their districts. This suggests that while legislators see themselves as Delegates, they may rarely be forced to *act* like a Trustee because of the perceived (real or not) policy homogeneity among the constituents in their districts. With respect to the variables associated with legislator role, education, and chamber are all associated with legislators perceiving themselves more as Trustees. Being black is associated with a legislator feeling more like a Delegate when compared to being white. State-level fixed effects that legislator role is associated with the state in which a legislator was elected. Political party, minority party status, gerrymandering, ideological polarization, gender, years in office, term limits, and professional legislatures have no statistically significant relationships with legislator role.

Of the relationships associated with a legislator's perception of constituent policy conflict in their districts, political party, ideological polarization, age, and education are all associated with a legislator perceiving *less* constituent policy conflict in their districts. State-level fixed effects analyses suggest that a legislator perception of constituent policy conflict in their district is associated with the state in which a legislator was elected. Minority party status, gerrymandering, gender, chamber race, years in office, professional legislature status, and term limits have no statistically significant relationships with a legislator's perception of constituent policy conflict in their districts.

Interestingly, the relationship between gerrymandering and district conflict detection is counterintuitive. It is reasonable to expect that gerrymandering would be correlated with conflict detection in a legislator's district; the more gerrymandered the district, the more homogeneous the policy preferences in the district, and the less policy conflict that would occur within the district. Even in simple bivariate regressions, only one gerrymandering measure, the Reock gerrymandering scale, had a statistically significant relationship with district conflict detection. A 1.0 percent increase in gerrymandering score (a decrease in gerrymandering) is associated with a 13 percent increase in conflict detection in a legislator's district. The sign on this relationship is as expected, but the significance of this finding is reduced in light of the non-statistically significant findings for all of the other gerrymandering measures.

The frequency of use of a communication technology legislators use to communicate with their constituents is associated with increases in legislators' perception of themselves as a Delegate. Face-to-face communications are the most effective at making legislators perceive themselves as Delegates, but hardcopy letters, telephone calls, web page communications, text messages, mass-media press, mass-media television, and mass-media radio all play a statistically significant role. The legislator perceived importance of a communication technology when used to communicate with constituents is also associated with a legislator feeling more like a Delegate. Statistically significant relationships include the importance of hardcopy letters, telephone calls, face-to-face meetings, web pages, text messages, mass-media television, town-hall meetings, mass-media radio, and blogs.

With respect to the frequency of use of a communication technology when used to communicate with constituents, only hardcopy letters, mass-media press, YouTube™, town-hall meetings, and mass-media radio have relationships with a legislator's perception of conflict in their district. Hardcopy letters *reduce* a legislator's perception of conflict, while mass-media press, YouTube™, town-hall meetings, and mass-media radio all *increase* a legislator's perception of constituent policy conflict in their district. With respect to the importance of a communication technology when used to communicate with constituents, only face-to-face meetings and telephone calls are statistically significant, and both *reduce* a legislator's perception of conflict in their district.

The frequency of use and importance of communication technologies legislators use to communicate *with each other* also have statistically significant relationships with legislator perception of their role and with legislator perception of conflict in their district. As expected, the peer relationships are much weaker than the frequency of use and importance legislators assign to communication technology when used to communicate with their constituents. Constituent communications are much more closely linked with legislator role and perception of constituent policy conflict in their district than are peer communications.

Since peer-to-peer communications are not typically considered linkages with constituents, they will not be discussed in detail except for one interesting finding. With respect to the importance of communication technologies to communicate *both* with constituents and peers, telephone calls and face-to-face meetings reduce legislators' perception of conflict in their districts, and *both* have similar coefficients. Face-to-face meetings with both peers and constituents cause a legislator to have approximately a 2.0 percent decrease in the perception of conflict in their district, while telephone calls to peers and constituents cause a legislator to have approximately a 1.0 percent decrease in the perception of conflict in their district.

Many of the findings in this chapter are novel from the perspective of existing research on legislator roles and perception of policy conflicts among their constituents. From this perspective, the results presented in this chapter may spur researchers into viewing legislator roles through a different lens. The bivariate relationships explored in this chapter, while statistically significant, have little explanatory power in and of themselves; the highest r-squared value for any bivariate regression presented in this chapter was 1.74 percent.

Locating the results of this chapter into the critical frequency theory suggests additional factors that influence the communication technologies that legislators use and value. Whether legislators view themselves as Delegates or Trustees and the amount of policy conflict in their district are all correlated with their choice of which communication technologies to use and value. Additionally, multivariate regression models highlight the role that DPI variables correlated with a legislator's perception of their role as a Delegate or Trustee, and correlated with a legislator's detection of policy conflict in their district. These factors highlight the complexity of feedback path attenuation associated with a legislator's role and district policy conflict.

NOTES

1. House of Representatives or Senate.
2. The Politico scale question asked legislators to reflect on Delegate/Trustee roles and not about specific instances of legislation, suggesting that legislator responses were an average value based on a legislator's perception of Delegate/Trustee roles.

Chapter 7

Political Polarization

OVERVIEW

The topic of ideological polarization in today's political arena is one of the most discussed aspects of political behavior, both with respect to legislators and their constituents. This chapter examines two potential sources of legislator political polarization: (1) DPI effects and (2) communication technology–related effects. As with chapter 6, this chapter deviates slightly from previous chapters insofar as the focus is not specifically on how legislators use and value communication technology as in previous chapters, but rather, this chapter is dedicated to examining the relationships between polarization and communication technology. Discussion begins with a brief overview of the literature surrounding political polarization and communication technology.

COMMUNICATION TECHNOLOGY AND POLITICAL POLARIZATION

The topic of political polarization and communication technology tends to center on the role that Internet-enabled communication technologies play in polarizing legislators and constituents. Some research suggests that Internet communications do not influence political polarization (Marshall, 2010), while others find that opinion reinforcement and challenge avoidance behaviors (selective exposure) can increase political polarization; individuals who have the same political ideology, when communicating without facing challenging opinions, end up taking more extreme ideological positions (Bimber and Davis, 2003; Garrett and Resnick, 2011; Sunstein and Sunstein, 2009). Although not directly related to communication technology, some of the most cited research on political polarization has been completed by Keith Poole and covers such topics as the relationship between gerrymandering and polarization (Poole, Rosenthal, and McCarty, 2009), and legislator polarization as reflected in legislator roll-call votes (Carroll and Poole, 2013; Poole, Lewis, Lo, and Carroll, 2008; Poole and Rosenthal, 2000; Poole and Rosenthal, 2011). With research mixed on whether or not communication technology use is associated with polarization, and no research located which directly examines the polarizing impact of communication technology on state legislators, this chapter, with bivariate and multivariate regression analyses using polarization as the dependent variable and various communication technologies and DPI independent variables, sheds additional light on this topic.

DEMOGRAPHIC, POLITICAL, AND INSTITUTIONAL POLARIZATION EFFECTS

Bivariate regressions were run to examine the relationships between political polarization measured on a five-point scale where legislators (Democrat, Republican, and Independent) were scored on their self-identified political party. For the ordinal *polarized* variable, legislators who identified as Independents were coded 1. Independent leaning Democrat and Independent leaning Republican were coded 2. Slightly progressive Democrats and slightly conservative Republicans were coded 3. Moderately progressive Democrats and moderately conservative Republicans were coded 4. Strongly progressive Democrats and strongly conservative Republicans were coded as 5. Additionally, a binary variable, *polarizedbin*, was created with a value of 1 for all legislators who self-identified as strongly conservative or strongly progressive, and 0 otherwise.

With respect to the bivariate regression results between the ordinal polarization measure and DPI variables, political ideology, education, the Reock gerrymandering scale, minority status in two chambers and the governorship, legislator Trustee/Delegate scale, and a legislator's perception of policy conflict in their district all have statistically significant relationships. The political ideology variable *partyid* is a measure of political conservativism with strong progressives (low level of conservativism) on the one end of the scale and strong conservatives on the other end of the scale. The relationship between polarization and conservative political ideology suggests that the more conservative a legislator, the more polarized they tend to be. The other statistically significant relationships are explained below.

A one-year increase in legislator education is associated with a 5.5 percent decrease in legislator polarization, on average, all else equal. With respect to legislator minority status (being in the minority party), legislators who are in the minority in both chambers and the governorship are 38.5 percent less likely to choose strongly progressive or strongly conservative polarization categories than their majority counterparts. A one-year increase in legislator education is associated with a 5.5 percent decrease in the probability that a legislator will choose strongly progressive or strongly conservative over all other polarization categories, on average, all else equal. A one unit increase in the Reock gerrymandering measure (less gerrymandering) is associated with a 2 percent increase in the probability that a legislator will select a strongly progressive or strongly conservative political ideology over all other political ideology categories, on average, all else equal. Increases in a legislator's perception of constituent policy conflict in their district and increases in legislator Delegate behavior are both associated with small decreases in the probability that a legislator will choose strongly progressive or strongly conservative out of all polarization categories, on average, all else equal.

Next, logistic probit bivariate regressions were completed with the dependent variable *polarizedbin* and all DPI variables. The intent of this analysis is to determine if there were any statistically significant relationships between DPI variables and the most polarized legislators. The DPI variables were then examined using a post-processing marginal effects dy/dx analysis. The results of these probit bivariate regressions are discussed below.

Of the DPI variables examined, political party, race (the "Other" political party), minority party status in both chambers and the governorship, and the conflict scale were statistically significant. Legislators who were in the minority party in both chambers and the governorship were 32 percent more likely to identify as strongly progressive Democrats or strongly conservative Republicans (in other words, at the extreme end of progressivism and conservativism), compared to all other polarization categories, on average, all else equal. Compared to white legislators, legislators who are not white and not black were 5 percent more likely to identify as strongly progressive or strongly conservative, compared to all other polarization categories, on average, all else equal. In the final statistically significant relationship, a 1 percent increase in the percentage of constituent policy conflict a legislator detects in their district is associated with a 0.2 percent increase in the probability that a legislator will identify as strongly progressive or strongly progressive compared to all other polarization categories, on average, all else equal.

With the DPI variables associated with polarization examined, it is time to turn our attention to the relationships between communication technology and polarization, first using bivariate linear regressions and then using bivariate logistic probit regressions followed by marginal effects analyses.

COMMUNICATION TECHNOLOGY POLARIZATION EFFECTS

Bivariate linear regressions were completed between the ordinal polarization variable described in the previous section and the importance and frequency of constituent and peer communication technologies when used to communicate with peers and constituents. The results of these bivariate linear regressions are discussed below.

With respect to the results associated with the polarization variable and the various communication technologies examined, a majority of the r-squared values are low, suggesting that while there are statistically significant relationships, the amount of variation in legislator polarization explained by variations in communication technology frequency of use and importance is low. This said, the largest effect sizes are associated with the importance of peer and constituent face-to-face communications, but not the frequency of use of peer and constituent face-to-face communications. Of the Internet-enabled communication technologies, the largest effect size occurs with peer frequency of use and importance of e-mails. With respect to the importance of constituent face-to-face meetings, a one category increase in the importance of face-to-face meetings with constituents is associated with a 24 percent increase in the probability that a legislator will identify as strongly progressive or strongly conservative over all other polarization categories, on average, all else equal. With respect to the frequency of e-mail use with constituents, a one category increase in the frequency of e-mails with constituents is associated with an 8.6 percent increase in the probability that a legislator will identify as strongly progressive or strongly conservative over all other polarization categories, on average, all else equal.

The final analyses in this chapter examine the statistically significant communication technology correlates with legislators who identify as strongly progressive or strongly conservative. The results of these analyses are discussed below.

Of the frequency of use and importance of the communication technologies examined, the following communication technologies were statistically significantly related to legislator polarization. With respect to the importance of communicating with constituents, face-to-face meetings, phone calls, hard-copy letters, e-mail, and Facebook™ were statistically significantly related to legislator polarization. With respect to the importance of communicating with peers, face-to-face meetings, phone calls, e-mail, Facebook™, and blogs were statistically significantly related to legislator polarization. With respect to the frequency of use of communication technologies to communicate with constituents, e-mail and Twitter™ use were statistically significantly related to legislator polarization. Finally, with respect to the frequency of use of communication technologies to communicate with peers, phone calls, e-mail, and

Facebook™ were statistically significantly related to legislator polarization. These results are discussed in greater detail in the following sections.

Importance of Communicating with Constituents and Legislator Polarization

Bivariate regressions were completed which examined the relationships surrounding the importance of communication technology when used to communicate with constituents and polarization. Statistically significant relationships were found to exist between polarization and the importance of face-to-face communications, phone calls, hardcopy letters, e-mails, and Facebook™ use when used to communicate with constituents. These regression analyses can be interpreted as follows: A one unit increase in the importance legislators assign to communicating with constituents via face-to-face meetings is associated with a 5.4 percent increase in the probability that a legislator will identify as strongly progressive or strongly conservative, on average, all else equal. A one unit increase in the importance legislators assign to communicating with constituents via phone calls is associated with a 2.9 percent increase in the probability that a legislator will identify as strongly progressive or strongly conservative, on average, all else equal. A one unit increase in the importance legislators assign to communicating with constituents via hardcopy letters is associated with a 3.4 percent increase in the probability that a legislator will identify as strongly progressive or strongly conservative, on average, all else equal. A one unit increase in the importance legislators assign to communicating with constituents via e-mail is associated with a 1.5 percent increase in the probability that a legislator will identify as strongly progressive or strongly conservative, on average, all else equal. A one unit increase in the importance legislators assign to communicating with constituents via Facebook™ is associated with a 1.3 percent increase in the probability that a legislator will identify as strongly progressive or strongly conservative, on average, all else equal. The fact that *every* mature communication technology is statistically linked to legislator polarization should not be lost on the reader; mature communication technologies are the most important to legislators, and it stands to reason that they would be linked to changes in political polarization.

Importance of Communicating with Peers and Legislator Polarization

Bivariate regressions were completed which examined the relationships surrounding the importance of communication technology when used to communicate with other legislators and polarization. Statistically significant relationships were found to exist between polarization and the importance of face-to-face communications, phone calls, e-mails, Facebook™, and blog use when used to communicate with other legislators. These regression analyses can be interpreted as follows: A one unit increase in the importance legislators assign to communicating with peers via face-to-face meetings is associated with a 5.5 percent increase in the probability that a legislator will identify as strongly progressive or strongly conservative, on average, all else equal. A one unit increase in the importance legislators assign to communicating with peers via phone calls is associated with a 3.5 percent increase in the probability that a legislator will identify as strongly progressive or strongly conservative, on average, all else equal. A one unit increase in the importance legislators assign to communicating with peers via e-mail is associated with a 1.5 percent increase in the probability that a legislator will identify as strongly progressive or strongly conservative, on average, all else equal. A one unit increase in the importance legislators assign to communicating with peers via Facebook™ is associated with a 1.4 percent increase in the probability that a legislator will identify as strongly progressive or strongly conservative, on average, all else equal. A one unit increase in the importance legislators assign to communicating with peers via blogs is associated with a 1.5 percent increase in the probability that a legislator will identify as strongly progressive or strongly conservative, on average, all else equal.

Frequency of Communicating with Constituents and Legislator Polarization

Bivariate regressions were completed which examined the relationships surrounding the importance of communication technology when used to communicate with other legislators and polarization. Statistically significant relationships were found to exist between polarization and the frequency of use of e-mails and Twitter™ when used to communicate with constituents. These regression analyses can be interpreted as follows: A one unit increase in the frequency that legislators communicate with constituents via e-mail is associated with a 1.2 percent increase in the probability that a legislator will identify as strongly progressive or strongly conservative, on average, all else equal. A one unit increase in the frequency that legislators communicate with constituents via Twitter™ is associated with a 1.2 percent increase in the probability that a legislator will identify as strongly progressive or strongly conservative, on average, all else equal.

Frequency of Communicating with Peers and Legislator Polarization

Bivariate regressions were completed which examined the relationships surrounding the frequency of use of communication technology when used to communicate with other legislators and polarization. Statistically significant relationships were found to exist between polarization and the frequency of use of phone calls, e-mails, and Facebook™ when used to communicate with constituents. These regression analyses can be interpreted as follows: A one unit increase in the frequency that legislators communicate with peers via phone calls is associated with a 2.1 percent increase in the probability that a legislator will identify as strongly progressive or strongly conservative, on average, all else equal. A one unit increase in the frequency that legislators communicate with peers via e-mail is associated with a 2.1 percent increase in the probability that a legislator will identify as strongly progressive or strongly conservative, on average, all else equal. A one unit increase in the frequency that legislators communicate with peers via Facebook™ is associated with a 1.6 percent increase in the probability that a legislator will identify as strongly progressive or strongly conservative, on average, all else equal.

Basic, Intermediate, and Advanced Regression Models

Three probit models (with dy/dx marginal effects post-processing) were created to determine the relationships between the polarization binary measure (a measure that *only* incorporates data from legislators who self-identified as strongly progressive or strongly conservative) and DPI variables. The basic model contains legislator demographic information, the intermediate model adds primary political and institutional variables, and the advanced model adds secondary institutional variables. Additionally, the results of these three models are not presented in a table, but rather, are discussed in the paragraphs following the models presented below:

Basic Model: $polarizedbin = \beta_0 + \beta_1(age) + \beta_2(race) + \beta_3(male) + \beta_4(education) + \mathcal{E}.$

Intermediate Model: $polarizedbin = \beta_0 + \beta_1(age) + \beta_2(race) + \beta_3(male) + \beta_4(education) + \beta_5(minority\ status) + \beta_6(years\ in\ office) + \beta_7(senate) + \mathcal{E}.$

Full Model: $polarizedbin = \beta_0 + \beta_1(age) + \beta_2(race) + \beta_3(male) + \beta_4(education) + \beta_5(minority\ status) + \beta_6(years\ in\ office) + \beta_7(senate) + \beta_8(gerrymandering) + \beta_9(professional\ legislature\ binary) + \beta_{10}(term\ limits) + \beta_{11}(conflict\ scale) + \beta_{12}(polarization\ binary) + \mathcal{E}.$

The basic model was statistically significant for age (dy/dx = −0.002, $z \leq 0.05$) and education (dy/dx = −0.018, $z \leq 0.05$). A one-year increase in legislator age is associated with a 0.2 percent *decrease* in the probability that a legislator will identify as strongly progressive or strongly conservative, on average, all else equal. A one-year increase in legislator education is associated with a 1.8 percent *decrease* in the probability that a legislator will identify as strongly progressive or strongly conservative, on average, all else equal. Putting these results in more practical terms, the older a legislator, and the more educated a legislator, the less likely they are to identify as strongly conservative or strongly progressive. The pseudo r-squared for the basic model is 1.25 percent.

The intermediate model was statistically significant for age (dy/dx = −0.009, $z \leq 0.01$), gender (dy/dx = −0.221, $z \leq 0.01$), education (dy/dx = −0.038, $z \leq 0.01$), and years in office (dy/dx = 0.016, $z \leq 0.01$). Being male results in a 2.2 percent *decrease* in the probability that a legislator will identify as strongly progressive or strongly conservative, on average, all else equal. A one-year increase in legislator education is associated with a 3.8 percent *decrease* in the probability that a legislator will identify as strongly progressive or strongly conservative, on average, all else equal. A one unit increase in legislator conservativism is associated with a 7.6 percent *decrease* in the probability that a legislator will identify as strongly progressive or strongly conservative, on average, all else equal. A one-year increase in legislator years in office is associated with a 1.6 percent *increase* in the probability that a legislator will identify as strongly progressive or strongly conservative, on average, all else equal.

Putting these results in more practical terms, male, older, more educated a legislator, the less likely they are to identify as strongly conservative or strongly progressive, while the longer a legislator is in office, the more likely they are to identify as strongly conservative or strongly progressive. The pseudo r-squared for the intermediate model is 4.4 percent.

The full model was statistically significant for age (dy/dx = −0.009, $z \leq 0.05$), gender (dy/dx = −0.215, $z \leq 0.05$), education (dy/dx = −0.04, $z \leq 0.01$), minority party status in both chambers and the governorship (dy/dx = −0.246, $z \leq 0.05$), and years in office (dy/dx = 0.018, $z \leq 0.01$). A one-year increase in legislator age is associated with a 0.9 percent *decrease* in the probability that a legislator will identify as strongly progressive or strongly conservative, on average, all else equal. Being male results in a 2.2 percent *decrease* in the probability that a legislator will identify as strongly progressive or strongly conservative, on average, all else equal. A one-year increase in legislator education is associated with a 4 percent *decrease* in the probability that a legislator will identify as strongly progressive or strongly conservative, on average, all else equal. Legislators who are in the minority party in both legislative chambers and the governorship are associated with a 21.5 percent *decrease* in the probability that a legislator will identify as strongly progressive or strongly conservative, on average, all else equal. A one-year increase in legislator years in office is associated with a 1.6 percent *increase* in the probability that a legislator will identify as strongly progressive or strongly conservative, on average, all else equal.

Putting these results in more practical terms, male, older, more conservative, more educated a legislator, the less likely they are to identify as strongly conservative or strongly progressive. Legislators who are in the minority party in both chambers and the governorship are less likely to identify as strongly conservative or strongly progressive compared to their majority party counterparts, and the longer a legislator is in office, the more likely they are to identify as strongly conservative or strongly progressive. The pseudo r-squared for the full model is 5.6 percent.

Full Regression Models

The previous sections analyzed and discussed three multivariate regression models associated with DPI variables. In this section, a legislator's self-identified polarization is examined as a function of communication technology frequency of use and importance.

$$polarizedbin = \beta_0 + \beta_1(pfreqmature) + \beta_2(pimportmature) + \beta_3(pfreqiect) + \beta_4(pimportiect) + \beta_5(cfreqmature) + \beta_6(cimportmature) + \beta_7(cfreqmass) + \beta_8(cimportmass) + \beta_9(cfreqiect) + \beta_{10}(cimportiect) + \mathcal{E}.$$

In this model, none of the communication technologies had statistically significant relationships with the binary polarization measure.

A second regression model, identical to the above model with the exception of the addition of all DPI variables:

$$Polarizedbin = \beta_0 + \beta_1(pfreqmature) + \beta_2(pimportmature) + \beta_3(pfreqiect) + \beta_4(pimportiect) + \beta_5(cfreqmature) + \beta_6(cimportmature) + \beta_7(cfreqmass) + \beta_8(cimportmass) + \beta_9(cfreqiect) + \beta_{10}(cimportiect) + \beta_{11}(age) + \beta_{12}(race) + \beta_{13}(male) + \beta_{14}(education) + \beta_{15}(minority\ status) + \beta_{16}(years\ in\ office) + \beta_{17}(senate) + \beta_{18}(gerrymandering) + \beta_{19}(professional\ legislature\ binary) + \beta_{20}(term\ limits) + \beta_{21}(conflict\ scale) + \beta_{22}(polarization\ binary) + \mathcal{E}.$$

In this model statistically significant relationships were found for the importance of Internet-enabled communication technologies when used to communicate with other legislators ($dy/dx = 0.296$, $z \leq 0.01$), education ($dy/dx = -0.019$, $z \leq 0.001$), minority party status in both chambers and the governorship ($dy/dx = -0.107$, $z \leq 0.01$), years in office ($dy/dx = 0.009$, $z \leq 0.01$), and the conflict scale ($dy/dx = -0.003$, $z \leq 0.01$). A one unit increase in the importance of using Internet-enabled communication technologies to communicate with other legislators is associated with a 29.6 percent increase in the probability that a legislator will identify as strongly progressive or strongly conservative, on average, all else equal. A one-year increase in legislator education is associated with a 1.9 percent *decrease* in the probability that a legislator will identify as strongly progressive or strongly conservative, on average, all else equal. Legislators who are in the minority party in both legislative chambers and the governorship are associated with a 33 percent *decrease* in the probability that a legislator will identify as strongly progressive or strongly conservative, on average, all else equal. A one-year increase in legislator years in office is associated with a 0.9 percent *increase* in the probability that a legislator will identify as strongly progressive or strongly conservative, on average, all else equal. A 1 percentage point increase in the amount of policy conflict a legislator detects in their district is associated with a 0.3 percent decrease in the probability that a legislator will identify as strongly progressive or strongly conservative, on average, all else equal.

Putting these results in more practical terms, more conservative, more educated legislators are less likely to identify as strongly conservative or strongly progressive. Legislators who are in the minority party in both chambers and the governorship are less likely to identify as strongly conservative or strongly progressive compared to their majority party counterparts, and the longer a legislator is in office, the more likely they are to identify as strongly conservative or strongly progressive. Increases in the importance of Internet-enabled communication technologies when used to communicate with other legislators are associated with increases in the probability a legislator will identify as strongly conservative or strongly progressive. The pseudo r-squared for this model is 10.1 percent.

Basic, Intermediate, and Advanced Regression Models

Three ordinal logistic regression models were created to determine the relationships between the *ordinal* (not the binary measure used in the models in the previous section) polarization measure and DPI variables. The results are presented in the form of odds ratios. The basic model contains legislator demographic information, the intermediate model adds primary political and institutional variables, and the advanced model adds secondary institutional variables. Additionally, the results of these three models are not presented in a table, but rather, are discussed in the paragraphs following the models presented below:

Basic Model: $polarization = \beta_0 + \beta_1(age) + \beta_2(race) + \beta_3(male) + \beta_4(education) + \mathcal{E}$.

Intermediate Model: $polarization = \beta_0 + \beta_1(age) + \beta_2(race) + \beta_3(male) + \beta_4(education) + \beta_5(minority\ status) + \beta_6(years\ in\ office) + \beta_7(senate) + \mathcal{E}$.

Full Model: $polarization = \beta_0 + \beta_1(age) + \beta_2(race) + \beta_3(male) + \beta_4(education) + \beta_5(minority\ status) + \beta_6(years\ in\ office) + \beta_7(senate) + \beta_8(gerrymandering) + \beta_9(professional\ legislature\ binary) + \beta_{10}(term\ limits) + \beta_{11}(conflict\ scale) + \mathcal{E}$.

The basic model was statistically significant for education (odds ratio = 0.943, $z \leq 0.001$). A one-year increase in legislator education is associated with a 5.8 percent *decrease* in the probability that a legislator will identify as polarized, on average, all else equal. Putting these results in more practical terms, the older a legislator, and the more educated a legislator, the less likely they are to identify as polarized. Although the coefficients are different, these results are similar to those obtained using the polarized binary variable. The pseudo r-squared for the basic model is 0.51 percent.

The intermediate model was statistically significant for gender (odds ratio = 0.654, $z \leq 0.001$), education (odds ratio = 0.963, $z \leq 0.05$), minority status in both chambers and the governorship (odds ratio = 0.756, $z \leq 0.05$), and years in office (odds ratio = 1.017, $z \leq 0.01$). Being male results in a 44.6 percent *decrease* in the probability that a legislator will identify as polarized, on average, all else equal. A one-year increase in legislator education is associated with a 3.7 percent *decrease* in the probability that a legislator will identify as polarized, on average, all else equal. Legislators in the minority party in both chambers and the governorship are associated with a 24.4 percent *decrease* in the probability that a legislator will identify as polarized, on average, all else equal. A one-year increase in legislator years in office is associated with a 1.6 percent *increase* in the probability that a legislator will identify as polarized, on average, all else equal. As with the basic model, these results are similar to the polarized binary variable results; however, they are more statistically significant and the coefficients tend to be larger.

Putting these results in more practical terms, legislators who are male, more conservative, more educated, and in the minority party in both legislative chambers and the governorship are less likely to identify as polarized, while the longer a legislator is in office, the more likely they are to identify as polarized. The pseudo r-squared for the intermediate model is 2.51 percent.

The full model was statistically significant for gender (odds ratio = 0.655, $z \leq 0.01$), education (odds ratio = 0.960, $z \leq 0.05$), minority party status in both chambers and the governorship (odds ratio = 0.641, $z \leq 0.01$), years in office (odds ratio = 1.016, $z \leq 0.05$), gerrymandering (odds ratio = 1.004, $z \leq 0.05$), delegate behavior (odds ratio = 0.996, $z \leq 0.01$), and conflict detection (odds ratio = 0.991, $z \leq 0.05$). Being male results in a 34.5 percent *decrease* in the probability that a legislator will identify as polarized, on average, all else equal. A one-year increase in legislator education is associated with a 4 percent *decrease* in the probability that a legislator will identify as polarized, on average, all else equal. Legislators who are in the minority party in both legislative chambers and the governorship are associated with a 35.9 percent *decrease* in the probability that a legislator will identify as polarized, on average, all else equal. A one-year increase in legislator years in office is associated with a 1.6 percent *increase* in the probability that a legislator will identify as polarized, on average, all else equal. A one unit increase in gerrymandering score (a decrease in gerrymandering) is associated with a 0.4 percent *increase* in the probability that a legislator will identify as polarized, on average, all else equal. A 1.0 percent increase in the amount of policy conflict a legislator detects in their district is associated with a 0.9 percent decrease in a legislator identifying as polarized. A 1.0 percent increase in a legislator acting like a Delegate is associated with a 0.4 percent increase in a legislator identifying as polarized.

Putting these results in more practical terms, being male, increases in education, increases in Delegate behavior, increased policy conflict detection in a district, and minority party status in both legislative chambers and the governorship are all associated with decreases in legislator polarization. Increases in conservative ideology, increases in years in office, and decreased gerrymandering are all associated with *increases* in legislator polarization. The pseudo r-squared for the full model is 5.7 percent.

Full Regression Models

The previous sections analyzed and discussed three ordinal multivariate regression models associated with DPI variables. In this section, a legislator's delegate score is examined as a function of communication technology frequency of use and importance.

$$polarization = \beta_0 + \beta_1 (pfreqmature) + \beta_2 (pimportmature) + \beta_3 (pfreqiect) + \beta_4 (pimportiect) + \beta_5 (cfreqmature) + \beta_6 (cimportmature) + \beta_7 (cfreqmass) + \beta_8 (cimportmass) + \beta_9 (cfreqiect) + \beta_{10} (cimportiect) + \varepsilon.$$

In this model, only the relationship with the importance of mature communication technologies when used to communicate with constituents was statistically significant (odds ratio = 0.874, $z \leq 0.001$). An increase in the importance of mature communication technologies when used to communicate with constituents is associated with a 22.6 percent increase in the probability that a legislator will identify as polarized, on average, all else equal. Put another way, polarization is associated with increases in the importance of communicating with constituents via face-to-face meetings, telephone calls, and hardcopy letters.

A second regression model, identical to the above model with the exception of the addition of all DPI variables:

$$polarization = \beta_0 + \beta_1 (pfreqmature) + \beta_2 (pimportmature) + \beta_3 (pfreqiect) + \beta_4 (pimportiect) + \beta_5 (cfreqmature) + \beta_6 (cimportmature) + \beta_7 (cfreqmass) + \beta_8 (cimportmass) + \beta_9 (cfreqiect) + \beta_{10} (cimportiect) + \beta_{11}$$

(age) + β_{12}*(race)* + β_{13}*(male)* + β_{14}*(education)* + β_{15}*(minority status)* + β_{16} *(years in office)* + β_{17}*(senate)* + β_{18}*(gerrymandering)* + β_{19}*(professional legislature binary)* + β_{20}*(term limits)* + β_{21}*(conflict scale)* + β_{22}*(polarization binary)* + $\mathcal{E}$.

The full model was statistically significant for the importance of Internet-enabled communication technologies when used with peers (odds ratio = 1.518, $z \leq 0.01$), the frequency of mature communication technology use to communicate with constituents (odds ratio = 0.707, $z \leq 0.01$), education (odds ratio = 0.942, $z \leq 0.01$), $z \leq 0.001$), minority party status in both chambers and the governorship (odds ratio = 0.603, $z \leq 0.01$), years in office (odds ratio = 1.032, $z \leq 0.01$), delegate behavior (odds ratio = 0.996, $z \leq 0.01$), and conflict detection (odds ratio = 0.986, $z \leq 0.01$). A one unit increase in the importance of Internet-enabled communication technologies when used to communicate with peers is associated with a 51.8 percent *increase* in the probability that a legislator will identify as polarized, on average, all else equal. A one unit increase in the frequency of use of mature communication technologies when used to communicate with constituents is associated with a 29.3 percent *decrease* in the probability that a legislator will identify as polarized, on average, all else equal. A one-year increase in legislator education is associated with a 5.8 percent *decrease* in the probability that a legislator will identify as polarized, on average, all else equal. Legislators who are in the minority party in both legislative chambers and the governorship are associated with a 39.7 percent *decrease* in the probability that a legislator will identify as polarized, on average, all else equal. A one-year increase in legislator years in office is associated with a 3.2 percent *increase* in the probability that a legislator will identify as polarized, on average, all else equal. A 1.0 percent increase in the amount of policy conflict a legislator detects in their district is associated with a 1.4 percent decrease in a legislator identifying as polarized. A 1.0 percent increase in a legislator acting like a Delegate is associated with a 0.4 percent increase in a legislator identifying as polarized.

Putting these results in more practical terms, increases in education, increases in Delegate behavior, increased policy conflict detection in a district, and minority party status in both legislative chambers and the governorship are all associated with decreases in legislator polarization. Increases in conservative ideology, increases in years in office, increases in the frequency of use of mature communication technology to communicate with constituents, and decreased gerrymandering are all associated with increases in legislator polarization. The pseudo r-squared for the full model is 6.2 percent.

CONCLUSION

Bivariate and multivariate regression analyses suggest the complexity of polarization in state legislators. As reflected by low pseudo r-squared values in the ordinal logistic and probit/marginal effects regressions performed in this chapter, the amount of variation in legislator polarization explained by variations in the use and importance of communication technology is relatively low. This said, the demographic, political, institutional, and communication technology variables examined in this chapter offer evidence that communication technology plays a small, but statistically significant role in explaining legislator polarization, with some communication technologies such as mature communication technology frequency of use with constituents being associated with decreases in polarization while other communication technologies, such as the importance of Internet-enabled communication technology use with peers being associated with increases in polarization.

When communicating with constituents, the frequency of use of Twitter™ and e-mail were statistically significantly correlated with legislator polarization; however, when communicating with other legislators, phone calls, e-mail, and Facebook™ were statistically significantly correlated with legislator polarization. When it comes to the importance of a communication technology when used to communicate with constituents, mature communications topped the polarization effects list, followed by the importance of e-mail and Facebook™. Interestingly, with the exception of the importance of hardcopy letter use (which legislators do not typically use to communicate with each other), mature communication technologies once again topped the polarization effects list, followed again by e-mail and Facebook™. In bivariate regressions, mature communication technologies, e-mail, and Facebook™ seem to be closely related to legislator self-identified polarized political ideology, whether the communications are with peers or constituents.

While ideological polarization does not directly address the role that communication technology plays in the critical frequency theory of policy system stability, it does hint at the complexity of the relationships between legislator behavior and communication technologies. As other researchers have discovered, the relationship between ideological polarization and communication technology is mixed; some communication technologies are correlated with ideological polarization, while others are not. The topic of political polarization wraps up the presentation of qualitative and quantitative results, and focus now turns toward examining the critical frequency theory of policy system stability in light of the analytical findings presented in chapters 2 through 7.

Chapter 8

Critical Frequency Theory of Policy System Stability

From a policymaking perspective, every theoretical model of the policy process in the United States has policy feedback as part of the model, either explicitly or implicitly. The critical frequency theory of policy system stability suggests that once the communication frequency (in communication events/unit time) reaches some institutional critical frequency ω_c (also called the phase crossover frequency), the policymaking system subjected to the communication events becomes unstable. This instability causes the policymaking system to speed up the process of creating and passing policies related to the disturbance. This chapter focuses on the role that communication technology and the DPI variables play in increasing or suppressing the frequency of communications from constituents which ultimately precipitate policy disturbances that destabilize policymaking systems and speed up the policymaking process.

Previous chapters have focused on communication technology frequency of use and importance, and the statistically significant (or not) DPI variables associated with communication technology frequency of use and importance. In this chapter, only the statistically significant relationships will be examined in light of the theory introduced in chapter 1. The goal of this chapter is to apply the concepts developed in chapter 1 to the results obtained in chapters 2 through 7 in order to clarify how communication technology impacts the policy process through the critical frequency theory of policy system stability.

Remembering back to chapter 1, there are four categories of feedback loop characteristics of particular interest when evaluating the potential impact of communication technology on the policy process. These four categories are: (1) feedback frequency, (2) crossover phase, (3) feedback path attenuation, and (4) feedback duration. Crossover phase (a characteristic of the policymaking institution) and feedback durations were not measured as part of the data collection for this book, leaving feedback path attenuation and feedback frequency as the primary focus for this chapter. Feedback frequency contributes to policy system *instability* by increasing the frequency of policy-related communication events closer to the critical frequency of the policymaking institution. Feedback path attenuation contributes to policy system *stability* by decreasing the number of policy-related communication through communication technology mismatches associated with DPI variables and perhaps more importantly through communication technology use and importance mismatches between legislators and constituents and between legislators and their peers.

Feedback path attenuation and feedback frequency are related; feedback path attenuation reduces the frequency of feedback, but it cannot increase it. Another distinction between feedback frequency and attenuation is that the use of a particular communication technology over which to receive and transmit policy information (feedback frequency) is a conscious choice, both by legislators and by constituents. As the interviews with legislators in chapters 2 and 4 suggest, the choice of which communication technologies to use or not to use is often personal preference. Feedback path attenuation is, in general, not a personal choice. As defined here, feedback path attenuation is associated with DPI relationships surrounding communication technology use. Race is a perfect example. Legislators are unlikely to say "I don't use mature communication technologies because I am Black," but rather, it is likely that black legislators do not use mature communication technologies as frequently as white legislators due to reasons not associated with personal choice. For example, if black constituents in mostly black districts do not use mature communication technologies to communicate with their legislators, then the legislators will report that they do not use mature communication technologies with their constituents. The result is that there are statistically significant differences in white and black legislator use of mature communication technologies, but these differences are unlikely to be personal choice.

Another way to think about the differences between feedback path attenuation and communication technology frequency of use is that feedback path attenuation can only stabilize a policymaking system. A communication sent from a constituent to a legislator (or from one legislator to another) would contribute toward destabilizing the policymaking system if it were received. However, if the communication were not received (or found to be less important because of the communication technology format) then the policymaking system is moved toward stability, just as if the communication had never been sent. Unlike feedback path attenuation, feedback frequency can both stabilize and destabilize a policymaking system. For example, if a constituent chooses not to communicate with their legislator on a particular policy, then the act of not communicating stabilizes the policymaking system. If on the other hand, a constituent chooses to communicate with their legislator and the communication is received, then the act of communicating moves the policymaking system toward instability. With this brief introduction, discussion begins with feedback path attenuation.

FEEDBACK PATH ATTENUATION

Feedback path attenuation occurs whenever policy preferences are communicated by a constituent but not received by a legislator (and vice versa), or are received by a legislator (or constituent) via a communication technology that is undervalued by the legislator due to personal preferences or because of DPI effects. As an example of personal preference, during interviews, one legislator noted that he was "a face-to-face" kind of guy and that he did not much care for e-mail. With respect to demographic effects, analyses suggest that black legislators use Internet-enabled communication technologies more frequently to communicate with their constituents than do white legislators. While this race finding may be the result of recognized personal preferences, it may also be related to how the constituents of black legislators choose to communicate based on (among other possibilities) their socioeconomic conditions. The second example is much more nuanced than the first. In the first example, the legislator notes that he does not like e-mail, which, if carried to the extreme, could mean that this legislator does not use e-mail at all with constituents; making her one of 933 such legislators who responded to the survey.

Unused Communication Technologies

Communication technology mismatches between legislators and policy stakeholders in the policy system environment stabilize the policymaking process by effectively reducing the number of communication events which reach the policymaking system. One indication of communication mismatches is the number of legislators who do not use a certain communication technology. If a legislator does not use a certain communication technology, all communication events from the environment via these communication technologies will not reach the policymaking system. Table 8.1 contains a list of the number of legislators who indicated that they do not use one or more of the various communication technologies examined in this book.

Table 8.1 suggests that legislators more frequently use mature communication technologies to communicate with constituents, which is of no surprise

Table 8.1 Communication Technology Not Used

Do Not Use	*e-mail*	*Twitter™*	*Facebook™*	*YouTube™*	*Web Pages*	*Blogs*	*Text Messages*	*F2F*	*Phone*	*Letters*
Constituent Frequency	933	469	651	1,236	461	1225	221	7	10	26
Peer Frequency	10	945	529	686	1,206	181	1,190	9	24	264

Self-generated by author using Microsoft Word.

given that the average age of the legislators participating in the research for this book is 57.9 years. This finding suggests that there is a potential communication mismatch between older legislators and younger constituents, who tend to use social media, text messages, and, to a lesser extent, e-mail (Kaid, 2004; McNair, 2017). One potential impact of this finding is that younger US citizens may be less likely to precipitate a state-level disturbance, and therefore less likely to have their policy concerns addressed. This finding is supported by existing research which recognizes that younger citizens in the United States are less politically engaged than older citizens (Bennett and Segerberg, 2012; Bimber, 2003; Kaid, 2004). Are communication technology mismatches between younger and older legislators partially responsible? Based on the data gathered for this book, there can be little doubt that the answer to this question is yes.

Communication Technology Mismatches

One way to identify communication technology mismatches is to examine differences between the importance a legislator assigns to a communication technology and the frequency that the communication technology is used. In chapter 2, legislators indicate that their most important form of mature communication technology when used with constituents is face-to-face communications. However, face-to-face meetings with constituents are the least frequently used mature communication technology. In interviews, legislators indicate that face-to-face interactions with constituents have the greatest influence over their policy decisions, and they lament that constituents do not attempt to meet with them face to face. This is a perfect example of a communication technology mismatch; legislators prefer and value face-to-face meetings with their constituents, but constituents are choosing to communicate via e-mail.

If legislators prefer (find it important) to communicate with constituents in face-to-face meetings, but constituents prefer to use e-mail (which many legislators either do not use or let their administrative assistant use for them), then it is reasonable to expect that the communications will be less impactful (or have zero impact) than they might otherwise have. In the best case, a legislator recognizes the mismatch and takes steps to address it (such as paying more attention to e-mail). Worst case, the legislator does not recognize the mismatch and continues to miss important communications from their constituents. This effectively opens the feedback loop, which has a stabilizing impact on the policy process, but in turn, also eliminates constituent feedback, with significant implications for a representative democracy such as that in the United States.

Demographic, Political, and Institutional Effects

Age

Although a significant body of research exists which explores the relationship between age and the use of the Internet and (in general) computers, none of the existing research uncovered for this book examined the role that age played in the use of Internet-enabled communication technologies by legislators. The findings in chapter 4 of this book indicate that age plays a significant role in the use of Internet-enabled communication technologies by legislators, just as age plays a role in the use of Internet-enabled communication technologies by non-legislators (Czaja, Sharit, Ownby, Roth, and Nair, 2001; Morrell, Mayhorn, and Bennett, 2000; Schleife, 2006; Thayer and Ray, 2006). The policy implications of age-based use of Internet-enabled communication technologies are significant. In American society, legislators tend to be older, and these older legislators are not as comfortable with Internet-enabled communication technologies as younger legislators. This said, there is another aspect to Internet-enabled communication technologies that give legislators reason to avoid it; once information is posted to the Internet, it lives forever and is virtually impossible to remove.

One legislator made a point of indicating the ability of information to live on forever, and the negative impact it can have on a legislative career. Here is the exchange:

Interviewer: So let me ask you this: have you had an incident where your use of communication technology has negatively impacted you? So you mentioned a little bit, you know, if you kind of have a knee-jerk response and it goes out and it could be played over and over and over again. Have you personally experienced that? Anything that you've communicated that was misunderstood or something that got replayed a bunch, that you had regretted? Related to communication technology?
Respondent: Because I've only run in one election and I didn't have a record as a new guy, it was hard to go back and get some clips of things I had said, but I did have one, I can remember. Our church—I'm LDS—our church did a digital media campaign and they asked everyone to go out and kind of tell people why you're a Mormon. So there's something called I am a Mormon and it's a huge community of hundreds of thousands of members of the church who have gone out and said this is why I'm a Mormon, this is what I believe. And I did that at the request of my church; I was a leader of a congregation of Hispanics.

And in my three or four paragraphs, I described—and it's still out there, you can go read it if you like—I had made the statement that I felt that SP 1070 had been very hard on my Hispanic congregation and that we had lost a third of our people and it just didn't seem like a real humanitarian thing to do.

And if I had any opportunity to be helpful in removing from office the people that made that happen, that I would do so. Well, a year later I was asked to run against the guy that wrote SP 1070—he was the Senate President, who was recalled.
Interviewer: Right, Name Retracted?
Respondent: Name Retracted. And so I announced that I was going to run on Saturday—two days before he announced he was going to run in March last year, a year ago.

And my campaign was gearing up and we announced. . . . I'm going to say we announced at nine in the morning; the media started picking up immediately and at about nine thirty, I realized. . . . My campaign was supposed to have scrubbed everything they could see on the internet, to see if there was anything that could be used against me.
Interviewer: Right, so that you're aware of it, basically.
Respondent: Yeah. We didn't change anything, but just be aware of what might happen when the media gets . . . oh, here's a guy that's going to run against Name Retracted. Name Retracted, blah-blah-blah . . . oh, he's got this I'm a Mormon thing.

And so sure enough, Steve Lemons at The New Times found I am a Mormon and was reading through it and my campaign was supposed to have read it first, but they hadn't thoroughly read it, and saw the two sentences that said I'll do all in my power to remove Name Retracted from office, the author of SP 1070.

And I just thought that was inappropriate, to have on a church site, given that I was going to run for office, to take him out of office. And so we quickly went out and removed that, but Steve Lemons had already taken a screen shot and so my first day in the new cycle was Name Retracted, the new guy, went out and removed these ten words from his "I am a Mormon site," and why did he do that?

And so the whole first couple of days of my campaign, I was doing damage control from a digital media quote that kind of got me off on the wrong foot. Now we got behind it quickly and I learned from that.
Interviewer: How would you do it differently?
Respondent: I would have looked at that first, got it removed, cleansed it before anyone knew I was running, and it was a mistake.
Interviewer: Would you have removed it again, I mean, even after the fact, would you have still chosen to remove it?
Respondent: If I could have done it before we announced, I would have still removed it, but if I would have known that people would have seen before and after, I would not . . . I would just live with it and. . . . Because they got to see it anyway.

This exchange makes it clear; being cautious of a form of communication technology is associated with some risks familiar to legislators and other high-profile individuals that the average citizen does not need to be as worried about. The effects of an avoidance of Internet-enabled communication technologies show up as an age-related phenomenon in quantitative modeling, but disentangling the root cause of a relationship is not a strength of quantitative analyses; qualitative analyses add detail that a researcher cannot predict.

Race

Race was a consistent theme in stabilizing the policymaking system by suppressing constituent communications with legislators. When using mature communication technologies, black legislators communicate less with their constituents via mature communication technologies, yet black legislators find mature communication technology just as important as white legislators. Black legislators use Internet-enabled communication technology to communicate with their constituents more frequently than do white legislators ($p \leq 0.001$), and they also find Internet-enabled communication technologies more important than do white legislators ($p \leq 0.001$). These relationships have the effect of suppressing policy feedback from older constituents in their district who tend to use mature communication technologies, and amplifying feedback from younger constituents who tend to use Internet-enabled communication technologies. Feedback from older constituents is lost, stabilizing the policy process, while feedback from younger constituents is amplified, pushing the policy process toward instability. If the policy disturbance is related to a policy important to older constituents, their voice is lost, and the probability of a quick policy change decreased. As discussed earlier, statistically significant relationships associated with race are difficult to disentangle without more information. Unfortunately, no legislators of color were able to be interviewed for this book, so for now, the quantitative results will need to stand on their own.

Gender

Research that focuses on legislator gender and communication technology generally concentrates on the substance of the communications as they relate to policy preferences, and not on the differences in the frequency of use and importance of communication technology associated with communication technology. For example Thomas (1991) notes that female legislators focus on "traditional" female concerns such as health, education, and welfare. Other researchers note that male and female legislators are perceived to support different constituencies (Reingold, 1992), and female legislators, in general, tend to support female constituents (Mansbridge, 2003; Porter, 2015; Reingold, 1992) and focus on policy preferences associated with "women's issues" (Bratton and Haynie, 1999; Dittmar, 2015; Fiorina, Abrams, and Pope, 2011; Mezey, 2008; Thomas and Wilcox, 2014).

A majority of the studies which examine gender-based communication technology frequency of use differences focus on non-legislators and a "digital divide" (Anderson, Wu, Cho, and Schroeder, 2015; Korunka and Hoonakker, 2014; West and Corley, 2016; Wohlers and Bernier, 2016) which is associated with gender-based communication differences, although many of these gender-based differences have been overcome in the United States (Cooper, 2004; Garson, 2006; Richardson and Cooper, 2006; Warschauer, 2004). Some researchers note that certain gender-based differences in the use of Internet-enabled communication technologies are associated with technological infrastructure disparities between males and females (Akman and Mishra, 2010; Hindman, 2010). Since legislators in this study share a common IT infrastructure, external variables such as access to the Internet, Internet connection speed, access to e-mail, and availability of a computer are normalized and should, theoretically, cease to be a source of frequency of use variations. In essence, there should be no "digital divide" influencing frequency of use when legislators share a common IT infrastructure.

In the previous chapters, many gender-based differences in the frequency of use and importance of various communication technologies have been noted (and will not be repeated here). Many of these differences are noted in simple bivariate regression models, but the statistical significance disappears once more complex multivariate regression models are used. However, there are some interesting differences in the frequency of use of mature and Internet-enabled communication technologies. As shown in figure 8.1, male legislators use mature communication technologies ($p \leq 0.001$) more than female legislators, and female legislators (figure 8.2) use Internet-enabled communication technologies ($p \leq 0.001$), more than male legislators. These differences have the effect of stabilizing policymaking systems when constituents communicate with their male legislators via Internet-enabled communication technologies and destabilizing policymaking systems when constituents communicate with their male legislators via face-to-face and telephone communications. With respect to female legislators, the policymaking system is stabilized when constituents communicate with female legislators via mature communication technologies and destabilizing the policymaking system when constituents communicate with their female legislators via Internet-enabled communication technologies.

From a policymaking perspective, the communication technology–based differences between male and female legislators suggest that constituents (and to a lesser extent, other legislators) would be well served to understand the communication preferences of each individual legislator, regardless of gender. This said, in general terms, it will be helpful for constituents especially, to understand that male legislators tend to prefer mature communication technologies, while female legislators tend to prefer Internet-enabled communication technologies.

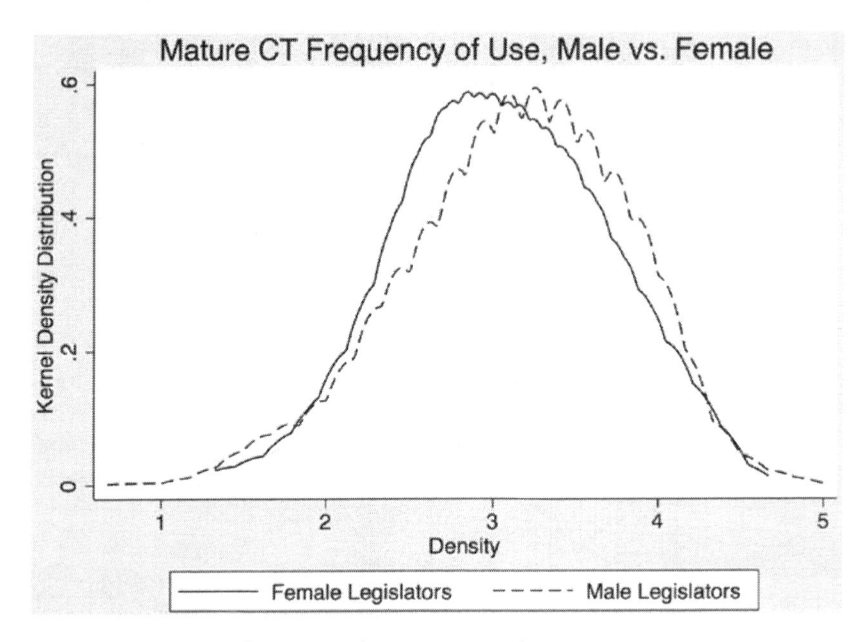

Figure 8.1 Mature CT frequency of use versus gender.
Self-generated by author using Stata Quantitative Analyses Program.

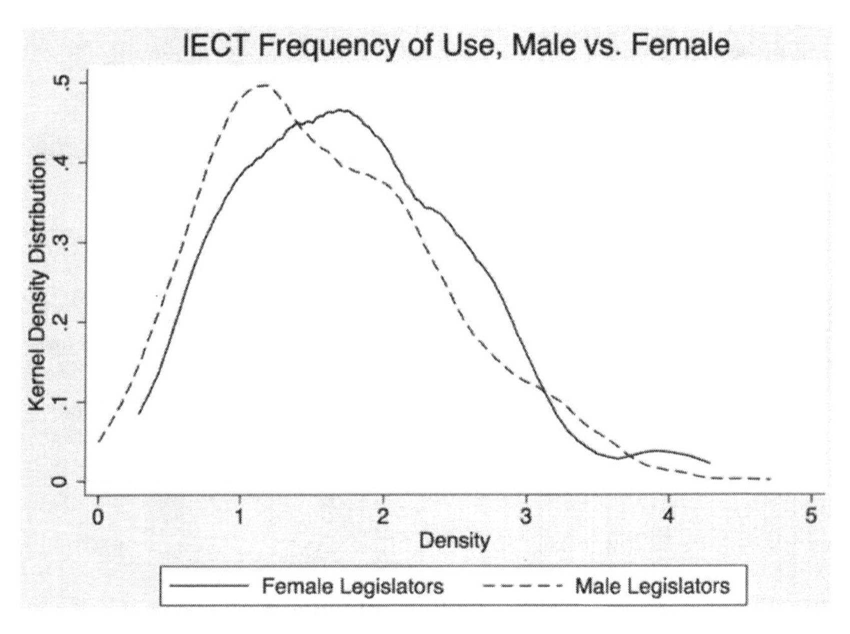

Figure 8.2 Internet-enabled CT frequency of use versus gender.
Self-generated by author using Stata Quantitative Analyses Program.

Means-testing (t-tests) was performed which examined the differences between male and female legislators with respect to the number of hours per week they meet with the following legislative stakeholders: (1) constituents, (2) staff, (3) lobbyists, (4) legislators in their own party, (5) legislators from other political parties, (6) constituents in their own party, (7) constituents from other parties, (8) legal staff, (9) administrators, and (10) executive elected officials. The results of these t-tests are contained in table 8.2.

Political Ideology

Political ideology also plays a role in stabilizing or destabilizing the policymaking process. The more conservative a legislator, the less likely they are to communicate with their constituents via Internet-enabled communication technologies, and the more likely they are to communicate with their constituents via mature communication technologies. These findings suggest that constituent communications with conservatives via Internet-enabled communication technologies stabilize the policymaking system and constituent communications via mature communications destabilize the policymaking system. These findings also suggest that constituent communications with progressive legislators via Internet-enabled communication technologies destabilize the policymaking system and constituent communications with progressive legislators via mature communications stabilize the policymaking system. Additionally, an increase in conservativism is associated with greater use of mature communication technologies, decreases in the use of press releases, radio and town-hall meetings, increases in the use of television to communicate with constituents, and decreases in the use of Internet-enabled communication technologies. The more conservative a legislator, the less important they find all Internet-enabled communication technologies except the use of web pages. Interestingly, the more conservative a legislator,

Table 8.2 Legislator Meeting Times versus Gender

Meetings With	*Male Mean (Hours Per Week)*	*Female Mean (Hours Per Week)*	*Statistical Significance*
Constituents	1.02	1.13	**
Legislative Staff	1.53	1.72	***
Lobbyists	1.12	1.18	
Legislators in Own Party	1.77	1.95	***
Legislators in Other Parties	1.20	1.41	***
Constituents in Own Party	0.74	0.82	**
Constituents in Other Parties	1.01	1.06	***
Legal Staff	0.70	0.72	
Administrators	0.97	1.10	**
Executive Elected Officials	0.82	0.93	***

Self-Generated by Author using Microsoft Word.

the *less* likely they are to detect policy conflict in their district. Whether this is because there is actually less conflict in conservative legislator districts or because conservatives are less aware of conflict in their district is impossible to ascertain with the data available for this book.

Education

The education level of a legislator also plays a statistically significant role in determining the stability of a policymaking system. Increases in legislator education level are associated with increases in the use of Internet-enabled communication technology to communicate with constituents, decreases in polarization, and decreases in the use of mature communication technologies to communicate with constituents and peers. (This includes the overall average of mature communications and face-to-face and hardcopy letter use individually, while telephone use decreased as a function of education.) These findings have the effect of destabilizing the policymaking system when constituents communicate with more highly educated legislators via Internet-enabled communication technologies and stabilizing the policymaking system when constituents communicate with more highly educated legislators via mature communication technologies. With respect to Internet-enabled communication technologies, Twitter™ and web page use with peers were both positively related increases in Internet-enabled communication technology frequency of use with peers, while e-mail, YouTube™, and web page use with constituents all increased with increasing education. One of the more interesting findings with respect to education is that the more educated a legislator, the more likely they are to act as a Trustee. This relationship passes face validity if one assumes that the more a legislator knows, the more likely they are to rely on what they know to make a policy decision.

Gerrymandering

As an example of an institutional effect, gerrymandering decreases policy feedback from constituents. With respect to mature and Internet-enabled communication technologies, when used with constituents, the lower the gerrymandering of districts, the higher the number of communication events. This relationship is true both with respect to bivariate regressions and multivariate regressions. These findings suggest that legislators communicate less with constituents the more gerrymandered the district the less legislators communicate with constituents. This in turn suggests that the more gerrymandered a legislator's district, the less of an impact the citizens in the district will have in creating a policy disturbance. Gerrymandering stabilizes the policy process by effectively suppressing constituent communications. With some of the important relationships with DPI variables covered, attention now turns to a discussion of each communication technology category.

MATURE COMMUNICATIONS

When communicating with their peers, legislators communicate most frequently via telephone conversations followed by face-to-face meetings, and finally by hardcopy letters, which are used on a monthly basis. With respect to the importance of mature communications with their peers, legislators find face-to-face meetings most important, followed by telephone calls and then hardcopy letters. When legislators communicate with their constituents, they use hardcopy letters most frequently, followed by telephone calls, and finally, face-to-face meetings most frequently. Legislators indicate that they prefer face-to-face meetings with their constituents, but they find that constituents do not engage them face-to-face. Legislators find face-to-face meetings most important for communicating with their constituents, followed by telephone use and finally, written letters.

Multivariate regression models on peer and constituent use and importance of mature communication technologies produce many statistically significant results. Among them are the following: (1) senators use all forms of mature communication technologies to communicate with other legislators more than Representatives; (2) the longer a legislator is in office, the more frequently they use mature communication technologies to communicate with their constituents, even after controlling for age and education; (3) the more polarized a legislator is, the more important they find face-to-face and telephone communications; and 4) black legislators use face-to-face communications with their peers less than white legislators, but they use telephone communications with their peers almost twice as frequently as their white counterparts. Kernel density distributions for both the frequency of use and importance of mature communications show that legislators more frequently use mature communications with constituents, and find these communications more important, when compared with peer communications.

MASS-MEDIA COMMUNICATIONS

With respect to other forms of communication technologies, legislators indicate that they use (mode = 0) and value (mode = 1) mass-media communication technologies significantly less than all other forms of communication technology examined in this book. A full comparison of each communication technology is contained in chapter 5. With respect to the relative frequency of use mass-media communication technologies, legislators use press releases most frequently (mode = 1), followed by television (mode = 0), town-hall meetings (mode = 0), and radio (mode = 0). With respect to the relative importance of mass-media communication technologies, legislators find press releases most important (mode = 3), followed by television (mode = 1), town-hall meetings (mode = 0), and finally radio (mode = 0). The following sections summarize the statistically significant relationships between the frequency of use and importance of mass-media communication technologies and various DPI variables.

Full mass-media communication technology frequency of use regression models for each type of mass-media communication technology suggest that compared to white legislators, black legislators use press releases and town-hall meetings more frequently, that one year of age reduces the use of town-hall meetings by 2 percent, television by 1 percent, and radio by 3 percent. For all mass-media communication technologies except radio, Senators use mass media more frequently than Representatives. The longer legislators are in office the more frequently they use town-hall meetings and television. The less gerrymandered a district, the more frequently legislators use all forms of mass-media communication technologies, while term limits decrease legislator use of town-hall meetings and television. Finally, the more conflict legislators detect in their district, the more likely they are to use town-hall meetings (or vice versa) to communicate with their constituents.

Full mass-media communication technology importance regression models for each type of mass-media communication technology suggest that as legislators age, they find all forms of mass-media communication technology less important. Male legislators find press releases less important than do female legislators. Black legislators find all forms of mass-media communication technology except press releases more important than white legislators. Senators find town-hall meetings and television more important than Representatives, while greater years in office are correlated with an increase in the perceived importance of town-hall meetings. The more gerrymandered a legislator's district, the higher the importance assigned to all forms of mass-media communication technology. Term limits are associated with decreases in the importance of town-hall meetings and television, while an increase in legislators' perception of constituent policy conflict.

INTERNET-ENABLED COMMUNICATIONS

There are several important themes which occurred in the analyses completed in chapter 4. First, with respect to legislator age, legislators use and value Internet-enabled communication technologies less the older they get. This phenomenon does not occur with the frequency of use of mature communication technologies. Second, even after controlling for the effects of age, the frequency of use and importance of Internet-enabled communication technologies when communicating with peers and constituents decreases the longer a legislator is in office. Third, male legislators use and value Internet-enabled communication technologies less than female legislators. Fourth, increases in legislator polarization are associated with both the frequency and importance of Internet-enabled communication technologies. Fifth, black legislators use and value Internet-enabled communication technologies more than their white counterparts. Sixth, increases in legislator Delegate behaviors are associated with increases in both the frequency and use of Internet-enabled communication technologies. Seventh, although the various gerrymandering measures did not provide consistent results, decreases in gerrymandering at the state level are statistically significantly correlated with increases in the frequency of use and importance of Internet-enabled communication technologies. The less safe a legislator's district is, the more likely they are to use and value Internet-enabled communication technologies.

OVERVIEW OF QUALITATIVE ANALYSES RESULTS

The qualitative research in chapter 6 uncovered legislator behaviors that were both expected and unexpected. Among the expected findings are the following: (1) legislators varied significantly in their use of communication technology; (2) legislators confirm the quantitative research on mature communication technology by verifying in their own words that more natural CTs are more important to them, primarily because natural CTs tend to convey information such as emotions and honesty/deception better than less natural CTs; and (3) demographic relationships between communication technology frequency of use and importance obtained during the quantitative research were confirmed. Significantly, communication technology frequency of use and importance rankings identified in the quantitative research were significantly correlated with the frequency of discussion for each communication technology discussed during qualitative interviews.

Network diagrams were created from qualitative coding and analyses, and were utilized to highlight the complexities surrounding communication technology, complexities that cannot be exposed by quantitative research. For example, qualitative research shows that legislators prefer not to use e-mail with constituents out of concern for FOIA laws, while quantitative research shows that e-mail frequency of use with constituents was ranked tenth out of fifteen, while e-mail frequency of use with peers was ranked number first. Put succinctly, quantitative research exposes the relationship, while qualitative research helps explain why the relationship exists.

More interesting than the expected findings were the unanticipated findings uncovered during qualitative research. These findings include the following: (1) there are significant mismatches between the communication technologies that legislators find most important and the communication technologies that constituents use most frequently (this was a common thread in the quantitative chapters); (2) legislators are using e-mail to avoid communications from constituents who disagree with their policy and/or political ideology and using e-mail to enhance communications with constituents who agree with their policy and/or political ideology; (3) legislators are using text messaging real time during floor debates to obtain outside information to assist them as they debate other legislators; (4) communication technology risk perceptions are related to the naturalness of the communication technology, with more natural communication technologies involving less perceived risk for the legislator; and (5) legislators use communication technologies as a challenge avoidance mechanism; legislators are more likely to engage with constituents with opposing ideologies if the engagement is face to face, slightly less likely to engage these constituents over the phone, and unlikely to engage these constituents via e-mail. Interviews suggest that naturalness theory offers an explanation for this behavior. Legislators believe they are unlikely to convince a constituent to change their mind about a topic via e-mail, slightly more likely to do so over the phone, and two legislators indicated they have changed the minds of every constituent they have ever met face to face.

Younger legislators are driving CT infrastructure development and changes, while older legislators are resisting these changes. In addition, technology itself is driving reductions in IT support staffing levels through automation made possible by new technologies. In an example of younger legislators driving changes in IT support infrastructure, several younger legislators requested IT support in setting up personal cell phones and computers to access the Arizona capitol intranet. These requests required secure tunneling VPN technology that was not in use by the IT department. The request was sent to the director and was approved, and now, setting up legislators' private computers and cell phones is one of the IT departments' most requested services. In another example, legislators requested bills be saved in PDF format in addition to Microsoft Word format.

Interviewee IT A categorized legislators into three fundamental categories: legislators who do not want to use technology at all, and have their legislator assistants deal with technology; legislators "in the middle" who use the technology but do not demand or embrace it; and finally, the third group of legislators who demand new communication technology. IT A indicated that older legislators tend to fall in the first category, while younger legislators tend to fall in the third category. IT B calls the first category of legislators "cowboy legislators" while noting that these older legislators wear cowboy boots under their suits.

Legislators are driving IT infrastructure changes and IT staff are adjusting their roles to support legislator requests. On the other hand, technology itself is driving reductions to IT staff by automating tasks that were once assigned to an IT staffer. IT A provides the example of web page creation. Web page creation was once a manual process and now it is automated through the use of scripts and custom programs. These factors suggest that IT infrastructure is a dynamic environment where technology can both add to the task load of IT personnel and reduce it. Interviews with IT staff suggest that legislators do in fact precipitate changes in IT infrastructure and support.

FINAL THOUGHTS

The research presented in this book explores a tremendous number of relationships between variables, so many relationships in fact, that it was sometimes difficult to determine what direction to take the analyses. There were times during the writing of this book that the analytical and theoretical choices available were overwhelming. One decision made early in the writing of this book was the choice to present all of the relevant results of the research, statistically significant or not. This decision was liberating because there was no longer any need to try and anticipate what other public policy theorists or political scientists might be interested in; there was sure to be sufficient evidence (significant or not) presented to satisfy at least a handful of researchers who study the confluence of communication technology and public policy. This said, there was so much more that could have been included in this book; the script file that was created to analyze the data used in this book is 4,654 lines long, and less than 15 percent of the results from this script file were used. Many more analyses were completed, both interesting and not so interesting, than could have possibly been included in this book given the space limitations imposed.

With respect to future research, one of the theories postulated in this book is that advances in communication technology should, all else equal, precipitate an increase in the number of policy system disturbances; the speed of digital communications and the ability to spread information more quickly and more widely than ever should cause faster issue expansion and result in an increase in the number of system disturbances. Do the availability and the relative speed and bandwidth of communication technology correlate to more policy system disturbances? The current issues of sexual assault and sexual harassment and the #MeToo movement highlight the speed and the disruption precipitated by Internet-enabled communication technologies, and in this case, Twitter™. The #MeToo movement (hashtag MeToo) caused many powerful men, including a number of legislators, to step down from their positions of power, both in the public and private sectors.

Any attempt to correlate communication technology advancement with policy system disturbances is fraught with dangers. Disturbances are complex. How would one disentangle the effects of communication technology from all the other variables associated with disturbances? Even if one could disentangle the impact of communication technology on the policymaking system, would it matter? Disturbances are inherently unpredictable and therefore any theory, including the venerable Punctuated Equilibrium Theory, on disturbances will not be useful for predicting disturbances (Baumgartner and Jones, 1993).

While it is relatively easy to argue that if all forms of communication to and from legislators were to cease, legislators would not be able to respond to environmental disturbances that precipitate policy disturbances, the argument is at best, a thought experiment. There is simply no realistic situation where this would occur. So, while communication technology as defined in this book is vital to the policymaking process, in realistic terms, the impact of communication technology on the policy process is difficult to disentangle from all of the other variables impacting the policy process. Based on the number of linear and nonlinear regression models presented in this book, the amount of variation in any dependent variable as a function of all of the independent variables available never exceeded 10 percent. This means that 90 percent of the variation in any dependent variable, including the frequency of use and importance of *any* communication technology, is unexplained. Simply put, DPI variables do not explain much of the variation in the use and importance of a communication technology. This said, the information in this book presents information that may open the door to further research, allowing others to stand on the shoulders of my work as I have stood on the shoulders of the work of the researchers listed in the bibliography for this book.

Does it matter that communication technology is contributing to instabilities in policymaking systems if there is no way to detect and use this information to predict policy instabilities? To answer this question, one needs only to consider technology itself. While, at the time of the writing of this book, it is difficult to conceive that it will be possible to monitor all of the communications to and from legislatures in order to measure their frequency and compare it to the critical frequency for the legislature, it is certainly *possible*, assuming that computing technology continues to advance. Even if we could tell when a policymaking institution was approaching instability, would we seek to use this information in some way that would benefit society?

Thinking back to the introduction to this book where we learned that legislators were acting as mouthpieces for lobbyists sitting in the gallery of a legislative chamber through the use of text messages, it is clear that communication technology is significantly impacting the policymaking process. The 10 percent of the variation in dependent variables explained by independent variables in the quantitative models presented in this book do not tell the whole story. Perhaps they do not even tell the most important story when it comes to how communicating technology is influencing the policy process. It is in fact possible that the true impact of communication technology on the policy process is related to the creative ways that policy stakeholders are using communication technologies in ways that state and federal constitutions never expected, nor could have sought to control. It is inconceivable that the drafters of the US Constitution could have conceived of a world where individuals could communicate across the world in milliseconds, yet here we are. Even though it is difficult to predict the communication technologies of the future even ten years from now, much less hundreds (or thousands) of years from now, it is fair, I think, to recognize that communication technology will play an important role in the policymaking processes of the future.

Appendix

Interviewee Descriptions

LEGISLATOR ASSISTANT DESCRIPTIONS

Assistant A: is a male assistant who works for two Representatives in the House. He is the youngest male legislator assistant interviewed and has less than five years of experience as a legislator assistant in the House.

Assistant B: is a female assistant who works for two Representatives in the House. She has more than twenty years of experience as a legislator assistant in the House. Assistant B works for legislator B.

Assistant C: is a female assistant who works for two Representatives in the House. She is middle-aged and has less than five years of experience as a legislator assistant in the House.

Assistant D: is a female assistant who works for one Senator in the Senate. She is an older assistant who has more than fifteen years of experience in both the House and Senate. Assistant D works for legislator D.

Assistant E: is a female assistant who works for two Representatives in the House. She is an older assistant and has more than fifteen years of experience as a legislator assistant in the House. Assistant E works for legislator E.

Assistant F: is a female assistant who works for two Representatives in the House. She is the youngest female assistant interviewed and has less than five years of experience as a legislator assistant in the House.

Assistant G: is a female assistant who works for one Senator in the Senate. She is an older assistant and has more than fifteen years of experience as a legislator assistant in both the House and Senate. Assistant G works for legislator G and belongs to the political party opposite the legislator she assists.

Assistant H: is a male assistant who works for one Senator in the Senate. He is the oldest male assistant interviewed, and has less than five years of experience as a legislator assistant in the Senate. Assistant H works for legislator H and belongs to the political party opposite of the legislator he assists.

Assistant I: is a female assistant who works for two Representatives in the House. She is the oldest female assistant interviewed and has more than fifteen years of experience as a legislator assistant in the House. Assistant I works for legislator I and belongs to the political party opposite the legislator she assists.

LEGISLATOR DESCRIPTIONS

Nine legislators were interviewed. They are identified as legislators B, D, E, G, H, I, J, K, and L. A brief description of each of these legislators is contained at the end of this section.

While legislator assistant and legislator questions are similar in nature (to aid in a comparison between legislator assistants and legislators), the understandings gained in previous interviews with legislator assistants help focus interactions with the legislator, allowing more time to be spent on exploring previously uncovered relationships rather than expending time and effort exploring basic relationships already uncovered during legislator assistant. For example, interviews with legislator assistants exposed how the assistant screens and filters each of these communication technologies for the legislator. During legislator interviews, the legislator was asked about their assistant's role in filtering and screening.

The methods used for legislator interviews were identical to the methods used for legislator assistants, and will not be repeated in this section. For instance, Krippendorff's alpha (α) was used to calculate intercoder agreement for legislator interviews, but it will not be discussed in detail since it was discussed at length in the legislator assistant instrument section.

The legislator interview protocol is broken into three sections comprised of non-directional open-ended questions. Closed-ended questions, where used, gate-sequenced closed-ended questions to improve the flow of the interview. Section one is comprised of an introduction and recording authorization. Section two, making up the main body of the interview, consists of sixteen interview questions spread across seven response categories that mirror the legislator assistant instrument. The section two response categories include communication technology frequency of use and preferences (four questions), the perceived risks and benefits of communication technology (two questions), behavior (two questions), roles and responsibilities (one question), new communication technologies (one question), constituents (four questions), and communication strategy (one question with five subparts).

The interview protocol concludes with an open-ended question, which asks the following: As an insider in the legislative process, what are some other important aspects of the relationship between legislators (yourself as well as other legislators) and communication technology that I have not touched upon in this interview? This final question is specifically designed to prompt the interviewee to reveal aspects of the relationships between legislators and communication technologies that were not covered in the previous sections. Section three contains closed-ended demographic questions, but these questions were not utilized as the information on legislators could easily be gathered from secondary sources.

Analysis of the legislator interview transcripts was completed by computer-aided qualitative analysis using the program ATLAS.ti. Qualitative research followed a grounded theory approach utilizing two cumulative coding cycles as outlined by Saldaña (2012) analytical memoing to generate networked relationships that were then analyzed for theory generation. Following a mixed methods approach, these results were compared and contrasted with the quantitative results.

LEGISLATOR DESCRIPTIONS

Legislator B: is a female Representative in her early sixties. She has less than five years of experience as a legislator.

Legislator D: is a female Senator in her early fifties. She has between five and ten years of experience as a legislator.

Legislator E: is a male Representative in his early sixties. He has between five and ten years of experience as a legislator.

Legislator G: is a male Senator in his early fifties. He has less than five years of experience as a legislator.

Legislator H: is a female Senator in her late fifties. She has between five and ten years of experience as a legislator.

Legislator I: is a male Representative in his mid-thirties and is the youngest Representative interviewed. He has less than five years of experience as a legislator.

Legislator J: is a female Representative in her early fifties. She has between five and ten years of experience as a legislator.

Legislator K: is a male Senator in his late sixties and is the oldest legislator interviewed. He has between ten and fifteen years of experience as a legislator.

Legislator L: is a male Senator in his late sixties. He has less than five years of experience as a legislator.

INFORMATION TECHNOLOGY STAFF DESCRIPTIONS

Three IT department professionals were interviewed. They are identified as IT A thorough IT C. Because the IT department is very small, no descriptions are provided for IT department professionals; any demographic or job description would likely be uniquely identifying, effectively removing anonymity for the interviewees.

References

Akman, I. and Mishra, A. (2010). Gender, Age and Income Differences in Internet Usage Among Employees in Organizations. *Computers in Human Behavior, 26*(3), 482–90.

Anderson, D., Wu, R., Cho, J. -S., and Schroeder, K. (2015). *E-Government Strategy, ICT and Innovation for Citizen Engagement.* Springer New York Heidelberg Dordrecht London: Springer.

Andrews, E. (2016). 10 Things You May Not Know About the Pony Express. Retrieved from http://www.history.com/news/history-lists/10-things-you-may-not-know-about-the-pony-express.

Arnold, R. (1992). *The Logic of Congressional Action.* Yale University Press.

Babbie, E. R. (2010). *The Basics of Social Research,* 5th Edition. Wadsworth Pub Co.

Baltar, F. and Brunet, I. (2012). Social Research 2.0: Virtual Snowball Sampling Method Using Facebook. *Internet Research, 22*(1), 57–74.

Baradat, L. P. and Philips, J. A. (2017). *Political Ideologies: Their Origins and Impact,* 12th Edition. New York: Routledge.

Bardach, E. (2016). *Practical Guide for Policy Analysis: The Eightfold Path to More Effective Problem Solving,* 5th edition. Sage.

Baum, M. A. and Groeling, T. (2008). New Media and the Polarization of American Political Discourse. *Political Communication, 25*(4), 345–65.

Baumgartner, F. and Jones, B. (1993). *Agendas and Instability in American Politics,* 2nd edition. Chicago, IL: University of Chicago Press.

Baumgartner, F. R. and Jones, B. (2002). *Policy Dynamics.* Chicago: University of Chicago Press, 3–28.

Beland, D. (2010). Reconsidering Policy Feedback. *Administration and Society, 42*(5), 568.

Bennett, W. L. and Segerberg, A. (2012). The Logic of Connective Action: Digital Media and the Personalization of Contentious Politics. *Information, Communication and Society, 15*(5), 739–68.

Bertalanffy, L. (1956). *General System Theory.* New York: George Braziller.

Bertot, J. C., Jaeger, P. T., and Grimes, J. M. (2010). Using ICTs to Create a Culture of Transparency: E-Government and Social Media as Openness and Anti-Corruption Tools for Societies. *Government Information Quarterly, 27*(3), 264–71.

Best, S. J. and Krueger, B. S. (2004). *Internet Data Collection.*

Bimber, B. A. (2003). *Information and American Democracy: Technology in the Evolution of Political Power.* Cambridge University Press.

Bimber, B. A. and Davis, R. (2003). *Campaigning Online: The Internet in US Elections.* USA: Oxford University Press.

Birkland, T. A. (2005). *An Introduction to the Policy Process: Theories, Concepts, and Models of Public Policy Making.* ME Sharpe Inc.

Bradley, R. B. (1980). Motivations in Legislative Information Use. *Legislative Studies Quarterly, 5*(3), 393–406.

Bratton, K. A. and Haynie, K. L. (1999). Agenda Setting and Legislative Success in State Legislatures: The Effects of Gender and Race. *The Journal of Politics, 61*(3), 658–79. doi:10.2307/2647822.

Bruns, A., Highfield, T., and Lind, R. A. (2012). Blogs, Twitter, and Breaking News: The Produsage of Citizen Journalism. *Produsing Theory in a Digital World: The Intersection of Audiences and Production in Contemporary Theory, 80*(2012), 15–32.

Burke, E. (1891). *The Works of the Right Honorable Edmund Burke . . . : Speeches and Correspondence.* G. Bell and sons.

Burke, K. and Chidambaram, L. (1999). How Much Bandwidth is Enough? A Longitudinal Examination of Media Characteristics and Group Outcomes. *MIS Quarterly, 23*(4), 557–79.

Butler, D. M. and Broockman, D. E. (2011). Do Politicians Racially Discriminate Against Constituents? A Field Experiment on State Legislators. *American Journal of Political Science, 55*(3), 463–77. doi:10.1111/j.1540-5907.2011.00515.x.

Caballer, A., Gracia, F., and Peiró, J. M. (2005). Affective Responses to Work Process and Outcomes in Virtual Teams: Effects of Communication Media and Time Pressure. *Journal of Managerial Psychology, 20*(3/4), 245–60.

Campbell, J. E. (1982). Cosponsoring Legislation in the US Congress. *Legislative Studies Quarterly, 7*(3), 415–22.

Campbell, K. (2003). *How Policies Make Citizens: Senior Political Activism and the American Welfare State.* Princeton University Press.

Canfield-Davis, K., Jain, S., Wattam, D., McMurtry, J., and Johnson, M. (2010). Factors of Influence on Legislative Decision Making: A Descriptive Study. *Journal of Legal, Ethical and Regulatory Issues, 13*(2), 415–22.

Carey, J. M., Niemi, R. G., Powell, L. W., and Moncrief, G. F. (2006). The Effects of Term Limits on State Legislatures: A New Survey of the 50 States. *Legislative Studies Quarterly, 31*(1), 105–34.

Carroll, R. and Poole, K. (2013). *Roll Call Analysis and the Study of Legislatures: The Oxford Handbook of Legislative Studies*, ed. Shane Martin, Thomas Saalfeld and Kaare Strøm. Oxford University Press.

Chen, C.-T. (1989). *System and Signal Analysis.* Harcourt School.

Chen, J. and Cottrell, D. (2016). Evaluating Partisan Gains from Congressional Gerrymandering: Using Computer Simulations to Estimate the Effect of Gerrymandering in the US House. *Electoral Studies, 44*, 329–40.

Chen, Y. and Persson, A. (2002). Internet Use Among Young and Older Adults: Relation to Psychological Well-Being. *Educational Gerontology, 28*(9), 731–44.

Cook, J. M. (2016). Twitter Adoption in US Legislatures: A Fifty-State Study. Paper presented at the Proceedings of the 7th 2016 International Conference on Social Media and Society.

Cook, T. E. (1989). *Making Laws and Making News: Media Strategies in the US House of Representatives*. Washington, DC: Brookings Institution Press.

Cooper, C. A. (2004). Internet Use in the State Legislature a Research Note. *Social Science Computer Review, 22*(3), 347–54.

Cox, G. W. and McCubbins, M. D. (2007). *Legislative Leviathan: Party Government in the House*. Cambridge University Press.

Creswell, J. W. and Clark, V. L. P. (2007). *Designing and Conducting Mixed Methods Research*. Wiley Online Library.

Cutler, S. J., Hendricks, J., and Guyer, A. (2003). Age Differences in Home Computer Availability and Use. *The Journals of Gerontology Series B: Psychological Sciences and Social Sciences, 58*(5), S271–S280.

Czaja, S. J., Sharit, J., Ownby, R., Roth, D. L., and Nair, S. (2001). Examining Age Differences in Performance of a Complex Information Search and Retrieval Task. *Psychology and aging, 16*(4), 564–79.

Daft, R. L. and Lengel, R. H. (1986). Organizational Information Requirements, Media Richness and Structural Design. *Management Science, 32*(5), 554–71.

Dahl, R. A. (1947). The Science of Public Administration: Three Problems. *Public Administration Review, 7*(1), 1–11.

Dahl, R. A. (1989). *Democracy and Its Critics*. Yale University Press.

Dahlgren, P. (2005). The Internet, Public Spheres, and Political Communication: Dispersion and Deliberation. *Political Communication, 22*(2), 147–62.

Davis, R. (1999). *The Web of Politics: The Internet's Impact on the American Political System*. USA: Oxford University Press.

Denhardt, R. (2008). *Theories of Public Organization,* 5th Edition. Belmont: Thomson Wadsworth.

Denhardt, R. B., Denhardt, J. V., and Aristigueta, M. P. (2008). *Managing Human Behavior in Public and Nonprofit Organizations*. Sage.

Dennis, A. R. and Kinney, S. T. (1998). Testing Media Richness Theory in the New Media: The Effects of Cues, Feedback, and Task Equivocality. *Information Systems Research, 9*(3), 256–74.

DeRosa, D. M., Hantula, D. A., Kock, N., and D'Arcy, J. (2004). Trust and Leadership in Virtual Teamwork: A Media Naturalness Perspective. *Human Resource Management, 43*(2–3), 219.

Dietrich, B. J., Lasley, S., Mondak, J. J., Remmel, M. L., and Turner, J. (2012). Personality and Legislative Politics: The Big Five Trait Dimensions Among US State Legislators. *Political Psychology, 33*(2), 195–210.

Dillman, D. A., Smyth, J. D., and Christian, L. M. (2009). *Internet, Mail, and Mixed-Mode Surveys: The Tailored Design Method,* 3rd edition. John Wiley and Sons Inc.
Dittmar, K. (2015). *Representation: The Case of Women.* Edited by Maria Escobar-Lemmon and Michelle M. Taylor-Robinson. New York: Oxford University Press, 2014.
Ellis, W. (2010). Committees, Subcommittees, and Information – Policymaking in Congressional Institutions. (PhD dissertation), University of Oklahoma, Oklahoma City.
Erikson, R. S. (1978). Constituency Opinion and Congressional Behavior: A Reexamination of the Miller-Stokes Representation Data. *American Journal of Political Science, 22*(3), 511–35.
Eulau, H., Wahlke, J. C., Buchanan, W., and Ferguson, L. C. (1959). The Role of the Representative: Some Empirical Observations on the Theory of Edmund Burke. *The American Political Science Review, 53*(3), 742–56.
Evans, H. K., Cordova, V., and Sipole, S. (2014). Twitter Style: An Analysis of How House Candidates Used Twitter in Their 2012 Campaigns. *PS: Political Science and Politics, 47*(02), 454–62.
Farwell, J. P. (2014). The media strategy of ISIS. *Survival, 56*(6), 49–55.
Fenno, R. F. (1973). *Congressmen in Committees*: Boston: Little Brown.
Ferber, P., Foltz, F., and Pugliese, R. (2005). Computer-Mediated Communication in the Arizona Legislature: Applying Media Richness Theory to Member and Staff Communication. *State and Local Government Review, 37*(2), 142–50.
Fiorina, M. P., Abrams, S. J., and Pope, J. C. (2011). *Culture War?* New York: Pearson Longman.
Forgette, R., Garner, A., and Winkle, J. (2009). Do Redistricting Principles and Practices Affect US State Legislative Electoral Competition? *State Politics and Policy Quarterly, 9*(2), 151–75.
Friedman, J. N. and Holden, R. T. (2009). The Rising Incumbent Reelection Rate: What's Gerrymandering Got to do with it? *The Journal of Politics, 71*(02), 593–611.
Gardner, J. A. (2006). Representation Without Party: Lessons from State Constitutional Attempts to Control Gerrymandering. *Rutgers Law Journal, 37,* 881.
Garrett, R. K. and Resnick, P. (2011). Resisting Political Fragmentation on the Internet. *Daedalus, 140*(4), 108–20.
Garson, G. D. (2006). *Public Information Technology and E-Governance: Managing the Virtual State.* Jones and Bartlett Learning.
Graber, D. A. (2009). *Mass Media and American Politics.* Sage.
Graham, T., Broersma, M., Hazelhoff, K., and van't Haar, G. (2013). Between Broadcasting Political Messages and Interacting with Voters: The Use of Twitter During the 2010 UK General Election Campaign. *Information, Communication and Society, 16*(5), 692–716.
Greenberg, S. R. (2012). *Congress and Social Media.* Retrieved from Center for Politics and Governance.
Gulati, G. J. and Williams, C. B. (2010). Congressional Candidates' Use of YouTube in 2008: Its frequency and Rationale. *Journal of Information Technology and Politics, 7*(2–3), 93–109.

Gulati, G. J. and Williams, C. B. (2013). Social Media and Campaign 2012 Developments and Trends for Facebook Adoption. *Social Science Computer Review, 31*(5), 577–88.

Hamajoda, A. F. (2016). Informing and Interacting with Citizens: A Strategic Communication Review of the Websites of the ECOWAS Parliaments. *The Electronic Journal of Information Systems in Developing Countries, 74*(1), 1–13.

Harden, J. (2013). Multidimensional Responsiveness: The Determinants of Legislators' Representational Priorities. *Legislative Studies Quarterly, 38*(2), 155–84.

Hayward, C. R. (2000). *De-facing Power*. Cambridge, UK; New York: Cambridge University Press.

Hedlund, R. and Friesema, H. (1972). Representatives' Perceptions of Constituency Opinion. *The Journal of Politics, 34*(03), 730–52.

Helfert, D. L. (2017). *Political Communication in Action: From Theory to Practice*. London: Lynne Rienner Publishers.

Herrick, R. and Thomas, S. (2005). Do Term Limits Make a Difference? Ambition and Motivations Among US State Legislators. *American Politics Research, 33*(5), 726–47. doi:10.1177/1532673x04270935.

Hindman, M. (2010). *The Myth of Digital Democracy*. Princeton University Press.

Huber, O. (1989). Information-Processing Operators in Decision Making. *Process and Structure in Human Decision Making*, 3–21.

Huurdeman, A. A. (2003). *The Worldwide History of Telecommunications*. New York: John Wiley.

Hwang, S. (2013). The Effect of Twitter Use on Politicians' Credibility and Attitudes Toward Politicians. *Journal of Public Relations Research, 25*(3), 246–58.

Ines, M. (2015). Opening Government: Designing Open Innovation Processes to Collaborate With External Problem Solvers. *Social Science Computer Review, 33*(5), 599–612.

Jewell, M. E. (1983). Legislator-Constituency Relations and the Representative Process. *Legislative Studies Quarterly, 8*(3), 303–37.

Jewell, M. E. and Patterson, S. C. (1966). *The Legislative Process in the United States*. Random House.

Jurich, J. A. and Myers-Bowman, K. S. (1998). Systems Theory and Its Application to Research on Human Sexuality. *The Journal of Sex Research, 35*(1), 72–87.

Juznic, P., Blazic, M., Mercun, T., Plestenjak, B., and Majcenovic, D. (2006). Who Says that Old Dogs Cannot Learn New Tricks?: A Survey of Internet/Web Usage Among Seniors. *New Library World, 107*(7/8), 332–45.

Kahneman, D. and Tversky, A. (1979). Prospect Theory: An Analysis of Decision Under Risk. *Econometrica: Journal of the Econometric Society, 47*(2) 263–91.

Kaid, L. L. (2004). *Handbook of Political Communication Research*. Routledge.

Kathlene, L. (1994). Power and Influence in State Legisllative Policy-Making: The Interaction of Gender and Position in Committee Hearing Debates. *American Political Science Review, 88*(3), 560–76. doi:10.2307/2944795.

Katz, D. and Kahn, R. L. (1978). *The Social Psychology of Organizations* (Vol. 2). New York: Wiley.

Keim, G. D. and Zeithaml, C. P. (1986). Corporate Political Strategy and Legislative Decision Making: A Review and Contingency Approach. *The Academy of Management Review, 11*(4), 828–43.

Khamis, S., Gold, P. B., and Vaughn, K. (2012). Beyond Egypt's " Facebook Revolution" and Syria's " YouTube Uprising:" Comparing Political Contexts, Actors and Communication Strategies. *Arab Media and Society, 15*, 1–30.

Kindra, G., Stapenhurst, F., and Pellizo, R. (2014). ICT and the Transformation of Political Communication. *International Journal of Advances in Management Science, 2*(1), 32–41.

Kingdon, J. (1984). *Agendas, Alternatives, and Public Policies* (Vol. 45). Boston: Little Brown.

Kingdon, J. and Thurber, J. (2003). *Agendas, Alternatives, and Public Policies*. New York: Longman White Plains.

Kingdon, J. W. (1989). *Congressmen's Voting Decisions*. University of Michigan Press.

Klain, R. A. (2007). Success Changes Nothing: The 2006 Election Results and the Undiminished Need for a Progressive Response to Political Gerrymandering. *Harvard Law and Policy Review, 1*, 75.

Knight, M. B. and Pearson, J. M. (2005). The Changing Demographics: The Diminishing Role of Age and Gender in Computer Usage. *Journal of Organizational and End User Computing (JOEUC), 17*(4), 49–65.

Kock, N. (2001). The Ape that Used E-Mail: Understanding E-Communication Behavior through Evolution Theory. *Communications of the Association for Information Systems, 5*(1), 3.

Kock, N. (2005). Media Richness or Media Naturalness? The Evolution of Our Biological Communication Apparatus and its Influence on Our Behavior Toward E-Communication Tools. *IEEE Transactions on Professional Communication, 48*(2), 117–30.

Kock, N. (2007). Media Naturalness and Compensatory Encoding: The Burden of Electronic Media Obstacles is on Senders. *Decision Support Systems, 44*(1), 175–87. doi:10.1016/j.dss.2007.03.011.

Korunka, C. and Hoonakker, P. (2014). *The Impact of ICT on Quality of Working Life*. Springer.

Larsson, A. O. and Moe, H. (2012). Studying Political Microblogging: Twitter Users in the 2010 Swedish Election Campaign. *New Media and Society, 14*(5), 729–47.

Lathrop, D. and Ruma, L. (2010). *Open Government: Collaboration, Transparency, and Participation in Practice*: Oreilly and Associates Inc.

Lawrence, N. W. (2011). *Social Research Methods: Qualitative and Quantitative Approaches,* 7th edition. Boston and London: Allyn and Bacon.

Lengel, R. H. and Daft, R. L. (1988). The Selection of Communication Media as an Executive Skill. *The Academy of Management Perspectives, 2*(3), 225–32.

Lessig, L. (2006). *Code: Version 2.0.* Basic Books.

Liikanen, J., Stoneman, P., and Toivanen, O. (2004). Intergenerational Effects in the Diffusion of New Technology: The Case of Mobile Phones. *International Journal of Industrial Organization, 22*(8–9), 1137–54.

Lilleker, D. G. and Koc-Michalska, K. (2013). Online Political Communication Strategies: MEPs, E-Representation, and Self-Representation. *Journal of Information Technology and Politics, 10*(2), 190–207.
Lindblom, C. and Woodhouse, E. (1993). *The Policy-Making Process,* 3rd edition. Prentice Hall.
Litton, N. (2012). Road to Better Redistricting: Empirical Analysis and State-Based Reforms to Counter Partisan Gerrymandering. *Ohio State Law Journal, 73*, 839.
Ludwig, V. B. (1972). The History and Status of General Systems Theory. *The Academy of Management Journal, 15*(4), 407–26.
Mansbridge, J. (2003). Rethinking Representation. *American Political Science Review, 97*(4), 515–28. doi:10.1017/S0003055403000856.
March, J. G. and Simon, H. A. (1958). *Organizations* (Vol. 1958). John Wiley and Sons.
Marshall, E. (2010). *Does the Internet Reinforce America's Partisan Divide?* (M.P.P), Georgetown University.
Maslow, A. (1943). A Theory of Human Motivation. *Psychological Review, 50*(4), 370–96.
Mayhew, D. R. (1974). *Congress: The Electoral Connection*. New Haven, CT: Yale University Press.
Mayhew, D. R. (2004). *Congress: The Electoral Connection* (Vol. 26). New Haven, CT: University Press.
McCarty, N. M., Poole, K. T., Rosenthal, H., and Knoedler, J. T. (2006). *Polarized America: The Dance of Ideology and Unequal Riches*. Cambridge, MA: MIT Press.
McCombs, M. E. and Shaw, D. L. (1972). The Agenda-Setting Function of Mass Media. *Public Opinion Quarterly, 36*(2), 176–87.
McNair, B. (2017). *An Introduction to Political Communication*. Taylor and Francis.
Mergel, I. (2010). Gov 2.0 Revisited: Social Media Strategies in the Public Sector. *PA Times, 33*(3), 7.
Mergel, I. (2012). *Social Media in the Public Sector: A Guide to Participation, Collaboration and Transparency in the Networked World*. Jossey-Bass.
Mezey, M. L. (2008). *Representative Democracy: Legislators and their Constituents*. Rowman and Littlefield Pub Inc.
Mettler, S. (1998). *Dividing Citizens: Gender and Federalism in New Deal Public Policy*. Ithaca, NY: Cornell University Press.
Miles, M. B., Huberman, A. M., and Saldaña, J. (2014). *Qualitative Data Analysis: A Methods Sourcebook,* 3rd Edition. Sage.
Miller, J. (2010). Utah Among 18 States Aiming to Copy Arizona Immigration Law. Retrieved from http://www.deseretnews.com/article/700043391/Utah-among-18-states-aiming-to-copy-Arizona-immigration-law.html.
Miller, W. E. and Stokes, D. E. (1963). Constituency Influence in Congress. *The American Political Science Review, 57*(1), 45–56.
Mills, C. W. (1956). *The Power Elite*. New York: Oxford University Press.
Mooney, C. Z. (1991). Information Sources in State Legislative Decision Making. *Legislative Studies Quarterly, 16*(3), 445–55.
Moran, C. T. (2013). Understanding Wisconsin Legislators' Use of Social Media. *University of Wisconsin Milwaukee UWM Digital Commons*, 1–102.

Morozov, E. (2011). *The Net Delusion: The Dark Side of Internet Freedom*. Public Affairs.

Morrell, R. W., Mayhorn, C. B., and Bennett, J. (2000). A Survey of World Wide Web Use in Middle-Aged and Older Adults. *Human Factors: The Journal of the Human Factors and Ergonomics Society, 42*(2), 175–82.

Mowrer, O. (1960). *Learning Theory and Behavior*. New York: Wiley.

Nahavandi, A. (2009). *The Art and Science of Leadership*. Pearson Prentice Hall.

Ogata, K. (1990). *Modern Control Engineering*. New Jersey: Prentice Hall.

Oleszek, W. J. (2007). *Congress and the Internet: Highlights*. Paper presented at the CRS Report for Congress.

Oleszek, W. J. (2011). *Congressional Procedures and the Policy Process*, 8th edition. CQ Press.

Overholser, A. (2016). An Examination of Sagebrush Rebellion Communications Using Narrative Policy Framework. *UNLV Theses, Dissertations, Professional Papers, and Capstones*.

Palfrey, J. G. and Gasser, U. (2013). *Born Digital: Understanding the First Generation of Digital Natives*. Basic Books.

Pasek, J., Kenski, K., Romer, D., and Jamieson, K. H. (2006). America's Youth and Community Engagement How Use of Mass Media is Related to Civic Activity and Political Awareness in 14- to 22-Year-Olds. *Communication Research, 33*(3), 115–35.

Patton, M. Q. (1990). *Qualitative Evaluation and Research Methods*. Sage.

Peterson, R. D. (2012). To Tweet or not to Tweet: Exploring the Determinants of Early Adoption Of Twitter by House Members in the 111th Congress. *Social Science Journal, 49*(4), 430–38. doi:10.1016/j.soscij.2012.07.002.

Pierson, P. (2000). Increasing Returns, Path Dependence, and the Study of Politics. *American Political Science Review, 94*(2), 251–67.

Poell, T. and Borra, E. (2012). Twitter, YouTube, and Flickr as Platforms of Alternative Journalism: The Social Media Account of the 2010 Toronto G20 Protests. *Journalism, 13*(6), 695–713.

Pole, A. J. (2005). E-Mocracy: Information Technology and the Vermont and New York State Legislatures. *State and Local Government Review, 37*(1), 7–24. doi:10.2307/4355383.

Poole, K., Lewis, J., Lo, J., and Carroll, R. (2008). Scaling Roll Call Votes with w-Nominate in R. *Journal of Statistical Software, 42*(14), 1–21.

Poole, K. T. and Rosenthal, H. (2000). *Congress: A Political-Economic History of Roll Call Voting*. USA: Oxford University Press.

Poole, K. T., Rosenthal, H., and McCarty, N. (2009). Does Gerrymandering Cause Polarization? *American Journal of Political Science, 53*(3), 666–80.

Poole, K. T. and Rosenthal, H. L. (2011). *Ideology and Congress* (Vol. 1). Transaction Books.

Porter, E. (2015). Women in Politics in Australia. *Women, Policy and Political Leadership, 1*, 73.

Proakis, J. G., Salehi, M., Zhou, N., and Li, X. (1994). *Communication Systems Engineering* (Vol. 2). New Jersey: Prentice Hall.

Reingold, B. (1992). Concepts of Representation among Female and Male State Legislators. *Legislative Studies Quarterly, 17*(4), 509–37. doi:10.2307/439864.
Richardson, G. P. (1991). *Feedback Thought in Social Science and Systems Theory*. University of Pennsylvania.
Richardson, L. E. and Cooper, C. A. (2006). E-mail Communication and the Policy Process in the State Legislature. *Policy Studies Journal, 34*(1), 113–29. doi:10.1111/j.1541-0072.2006.00148.x.
Rohr, J. (1986). *To Run a Constitution: The Legitimacy of the Administrative State*. Lawrence: University Press of Kansas.
Rosenthal, A. (1997). *The Decline of Representative Democracy: Process, Participation, and Power in State Legislatures*. CQ Press.
Rourke, L., Anderson, T., Garrison, D. R., and Archer, W. (1999). Assessing Social Presence in Asynchronous Text-Based Computer Conferencing. *Journal of Distance Education, 14*(2), 50–71.
Sabatier, P. (2007). *Theories of the Policy Process*, 2nd edition. Cambridge: Westview Press.
Sabatier, P. (2014). *Theories of the Policy Process*, 3rd edition. Cambridge: Westview Press.
Sabatier, P. and Jenkins-Smith, H. (1993). *Policy Change and Learning: An Advocacy Coalition Approach*. Westview Press.
Sabatier, P. and Weible, C. (2014). *Theories of the Policy Process*, 3rd edition. Westview Press.
Sala, J. F. A. and Jones, M. P. (2012). The Use of Electronic Technology and Legislative Representation in the Mexican and US States: Nuevo León and Texas. *Puentes Consortium Project*. Retrieved January, 17, 2016.
Saldaña, J. (2012). *The Coding Manual for Qualitative Researchers*. Sage.
Saldaña, J. (2015). *The Coding Manual for Qualitative Researchers*. Sage.
Schattschneider, E. (1975). *The Semi-Sovereign People: A Realist's View of Democracy in America*. New York: Dryden Press.
Scheflen, A. E. (1972). *Body Language and the Social Order; Communication as Behavioral Control*. Prentice-Hall, 208.
Scheufele, D. A. (2002). Examining Differential Gains from Mass Media and their Implications for Participatory Behavior. *Communication Research, 29*(1), 46–65.
Schleife, K. (2006). Computer Use and Employment Status of Older Workers—An Analysis Based on Individual Data. *Labour, 20*(2), 325–48.
Schmandt-Besserat, D. (1996). *How Writing Came About*. Austin: University of Texas Press.
Schneider, A. and Ingram, H. (1997). *Policy Design for Democracy*. Lawrence: University Press of Kansas.
Scott, W. and Davis, G. (2007). *Organizations and Organizing: Rational, Natural, and Open Systems Perspectives*. Prentice Hall.
Selnow, G. W. (1998). *Electronic Whistle-Stops: The Impact of the Internet on American Politics*. Praeger Publishers.
Selwyn, N., Gorard, S., Furlong, J., and Madden, L. (2003). Older Adults' Use of Information and Communications Technology in Everyday Life. *Ageing and Society, 23*, 561–82. doi:10.1017/s0144686x03001302.

Shapiro, A. (1999). *The Control Revolution: How the Internet is Putting Individuals in Charge and Changing the World We Know*. Public Affairs.

Shapiro, M. A. and Hemphill, L. (2016). Politicians and the Policy Agenda: Does Use of Twitter by the US Congress Direct New York Times Content? *Policy and Internet, 9*(1), 109–32.

Shirky, C. (2008). *Here Comes Everybody: The Power of Organizing Without Organizations*. Penguin Press.

Short, J., Williams, E., and Christie, B. (1976). The Social Psychology of Telecommunications. *Journal of Computer and Communications, 2*(14), 387–400.

Simon, H. (1957). *Administrative Behavior: A Study of Decision-Making Processes in Administrative Organizations*. Free Press.

Skinner, B. (1965). *Science and Human Behavior*. Free Press.

Smith, E. R. A. N. (1989). *The Unchanging American Voter*. University of California Press.

Soss, J., Hacker J., and Mettle, S. (2010). *Remaking America: Democracy and Public Policy in an Age of Inequality*. New York: Russell Sage.

Soss, J. and Schram, S. (2007). A Public Transformed? Welfare Reform as Policy Feedback. *American Political Science Review, 101*(01), 111–27.

Stewart, J. and Ayres, R. (2001). Systems Theory and Policy Practice: An Exploration. *Policy Sciences, 34*(1), 79–94.

Straus, J. R., Glassman, M. E., Shogan, C. J., and Smelcer, S. N. (2013). Communicating in 140 Characters or Less: Congressional Adoption of Twitter in the 111th Congress. *PS: Political Science and Politics, 46*(01), 60–66.

Sue, V. M. and Ritter, L. A. (2011). *Conducting Online Surveys*. Sage.

Sunstein, C. R. and Sunstein, C. R. (2009). *Going to Extremes: How Like Minds Unite and Divide*. Oxford: Oxford University Press.

Thayer, S. E. and Ray, S. (2006). Online Communication Preferences Across Age, Gender, and Duration of Internet Use. *CyberPsychology and Behavior, 9*(4), 432–40.

Thomas, S. and Welch, S. (1991). The Impact of Gender on Activities and Priorities of State Legislators. *The Western Political Quarterly, 44*(2), 445–56. doi:10.2307/448788.

Thomas, S. and Wilcox, C. (2014). *Women and Elective Office: Past, Present, and Future*. Oxford University Press.

Trevino, L. K., Lengel, R. H., and Daft, R. L. (1987). Media Symbolism, Media Richness, and Media Choice in Organizations a Symbolic Interactionist Perspective. *Communication Research, 14*(5), 553–74.

Tufekci, Z. and Wilson, C. (2012). Social Media and the Decision to Participate in Political Protest: Observations from Tahrir Square. *Journal of Communication, 62*(2), 363–79.

Von Bertalanffy. (1950). An Outline of General System Theory. *The British Journal for the Philosophy of Science, 1*(2), 134–65.

Wahlke, J. C. (1962). *The Legislative System: Explorations in Legislative Behavior*. Wiley.

Walgrave, S., Varone, F., and Dumont, P. (2006). Policy with or without Parties? A Comparative Analysis of Policy Priorities and Policy Change in Belgium, 1991 to 2000. *Journal of European Public Policy, 13*(7), 1021–38.

Walker, J. (1969). The Diffusion of Innovations Among the American States. *The American Political Science Review, 63*(3), 880–99.

Wang, S. and Feeney, M. K. (2016). Determinants of Information and Communication Technology Adoption in Municipalities. *The American Review of Public Administration, 46*(3), 292–313.

Warschauer, M. (2004). *Technology and Social Inclusion: Rethinking the Digital Divide*. Cambridge, MA: MIT Press.

West, J. (2014). *Evolutionary Changes: The Complex Relationships Between Legislators and Communication Technology.* (PhD dissertation), Arizona State University, Ann Arbor. ProQuest Dissertations and Theses Full Text database.

West, J. F. and Corley, E. A. (2016). An Exploration of State Legislator Communication Technology Use and Importance. *Journal of Information Technology and Politics, 13*(1), 52–71.

Wiener, M. and Mehrabian, A. (1968). *Language within Language: Immediacy, a Channel in Verbal Communication*. Irvington Pub.

Williams, C. B. and Gulati, G. J. (2009). Facebook Grows Up: An Empirical Assessment of Its Role in the 2008 Congressional Elections. *Proceedings from Midwest Political Science Association, Chicago.*

Williams, C. B., Gulati, J., and DeLeo, R. (2013). *The Dissemination of Social Media to Campaigns for State Legislature: The 2012 New England Case.* Paper presented at the APSA 2013 Annual Meeting Paper.

Wohlers, T. E. and Bernier, L. L. (2016). *Setting Sail into the Age of Digital Local Government* (Vol. 21). Christopher G. Reddick, San Antonio, USA: Springer.

Worsham, J. (2006). Up in Smoke: Mapping Subsystem Dynamics in Tobacco Policy. *Policy Studies Journal, 34*(3), 437–52.

Wyld, D. (2007). *The Blogging Revolution: Government in the Age of Web 2.0.* IBM Center for the Business of Government.

Wyrick, T. L. (1991). Management of Political Influence Gerrymandering in the 1980s. *American Politics Quarterly, 19*(4), 396–416.

Index

About the Author

Joe West is an electrical engineer who focused on satellite and launch systems communication technologies for more than thirty years, working for organizations such as Motorola Government Electronics Group, the National Aeronautics and Space Administration (NASA), Orbital Sciences Corporation, and Intel Corporation. After earning his PhD in public policy, West took an assistant professor position at the University of North Carolina at Pembroke where he teaches graduate-level quantitative analysis and research methods courses in the Master of Public Administration (MPA) and undergraduate political science programs. In his spare time, West is a PADI SCUBA instructor and a licensed private pilot who enjoys restoring antique automobiles, off-roading in his Jeep, competitive shooting, motorcycle riding, and running with scissors.